A. McDowell

# Instruction Book for Drafting and Cutting Dresses, Basques, Sacks, Coats, etc.

A. McDowell

**Instruction Book for Drafting and Cutting Dresses, Basques, Sacks, Coats, etc.**

ISBN/EAN: 9783337314224

Printed in Europe, USA, Canada, Australia, Japan

Cover: Foto ©berggeist007 / pixelio.de

More available books at **www.hansebooks.com**

—FOR—

# Drafting and Cutting

## DRESSES, BASQUES, SACKS, COATS, &C.

—BY THE—

## GARMENT DRAFTING MACHINE,

*AS INVENTED AND PATENTED*

BY

A. McDOWELL

---

**The McDowell Garment Drafting Machine Co.**

No. 6 West 14th Street, New York, U. S. A.

SEVENTH EDITION.

---

*NEW YORK:*

MERCANTILE PRINTING & STATIONERY CO., [illegible] BROADWAY, N. Y.

1885.

# Hints to Dressmakers.

---

## BASTING AND FINISHING.

One reason why foreign dresses are so much admired is the splendid manner in which they are finished.

Proper basting is a very important matter, for if the lining and the outside, although cut properly, are not united correctly the result will not be such as was intended.

Basting then is the foundation of making, and its importance should be generally appreciated in this country. Don't be afraid of using too many stitches.

### FIRST BASTE THE SEPARATE PIECES OF LINING TO THE OUTSIDE.

Place the material on the table with the wrong side up, with the wrinkles all nicely smoothed out; then lay the lining of the front on the goods and baste on the sewing lines, using a stitch about one inch long. Baste down the fold line evenly to two inches below the point of the bust, from this point to the waist full the lining in about one quarter of an inch. From the waist line down baste evenly. When basting around the darts, full the lining from one inch below the top of the darts to the waist line about one-quarter of an inch; continue from the waist down plain.

The goods and the lining are basted evenly under the arm, from the arm-hole down, except for the three inches just above the waist, the lining is to be fulled a little. The lining on the shoulder is fulled slightly. Lay the lining of the back and the side body properly on the material and baste in the sewing lines evenly except for the three inches above the waist line; the lining is to be slightly fulled in these three inches, say one-quarter inch. Below the waist baste evenly.

### NEXT BASTE THE SEAMS.

Begin one quarter of an inch above the top of the front darts and baste evenly to the waist line, and continue from there down. When the second dart slants very much, it will be necessary to full the edge next to the front slightly down as far as the waist. The hip dart is basted evenly from the armhole to the waist; from there down the edge next to the front is fulled slightly for three inches, then plain. In joining the side body to the back, begin at the armhole, baste evenly for two inches and then full the back slightly for the next three inches; from there down baste evenly. Always hold the back towards you when basting. Baste the side seam evenly, beginning at the armhole. In joining the front and back at the shoulder, full the back and stretch the seam after basting.

To prevent the armhole from stretching, use a strong thread or cord around it when basting. Having properly basted the garment, stitch the seam exactly on the sewing lines. This must be done very carefully. Press the front darts into shape by drawing them over the knee, then press all the seams and stay the edges, first trim slightly and then overseam the edges separately. For a better finish bind each edge inside with silk binding. All dresses should be finished with an inside belt, which must be fastened on each seam, thus relieving the goods of the waist strain. The casing for the whalebones in cheap dresses can be made out of the seams, but for good dresses it is best to press the seams open and then cover the whalebone with a piece of the lining, cut bias and nicely felled on, or with a piece of tape. The whalebone must be of good quality and thin at the ends. Clipping the seams is necessary, to have the goods lay smoothly. Never trim the shoulder and side seams too closely. For fleshy figures, particularly those that break open the seams, cut the lining crosswise when the material will permit, that is have the selvedge top and bottom. Some good cutters always cut their linings this way, but the majority cut the lining lengthwise, the same as the outside. Never allow a customer to put her thumb into the armhole when fitting the dress, as she will thereby stretch the armholes and cause a fullness in the front of the arm very hard to remedy.

Learn to make garments well, and induce your customers, if possible, to wear becoming styles and color. As a rule, rich trimming should accompany rich material, and colors should invariably harmonize.

Study these points, and when seeking a position as forelady, or cutter and fitter, your value will be greatly enhanced.

If you would become an expert, and we hope that such is your aim, you can command your own price.

## HOW TO TAKE THE MEASURE.

Remember that a perfect-fitting garment can only be produced by first securing a perfect measure. Before beginning to take a measure always place a belt of stiff material, two inches wide, tightly around the waist, and see that its lower edge is just at the bottom of the waist—not crowded down too much, nor pushed up by the skirts rolled on the band. Have the lower edge just at the natural waist.

Fig. 1.

FIRST.—The *neck* measure is taken, (while standing at the back) just above the band of the dress, around the bare neck, snug, not loose. Do not allow anything for comfort; the machine does that This measure is shown at 1 on Fig. 1.

SECOND.—*The width of back* is taken with the tape across the shoulder-blades, between the armholes, and is to be just the width you desire the garment to be when finished, as the machine allows for the seams. This measure is shown by the line from 2 to 3 on Fig. 1.

This measure can also be taken by using the square as shown on Fig. 4, from 21 to 22. To use the square, place its short end under the left arm, and slide up the *gauge* on the long part until it touches the body under the right arm. The figures on the square at the right or outer side of the gauge furthest from the body, give the correct measure. It is best to measure with the tape when you understand how.

THIRD.—*The under-arm length* is taken with the open hand resting on the hip, one inch below the belt, as shown in Fig. 2. Take the end of the tape between the thumb and first finger of the right hand, and pass it under the arm with the second joint of the thumb touching the arm, keeping the thumb and finger straight through, neither pointing up nor down. Then with the left hand bring the tape to the lower edge of the belt. Hold it there with the left hand and draw back the right hand, and if the knuckles just touch the arm, with the tape stretched, it will give the correct measure. This measure is shown on Fig 2, from 12 to 13. Always measure both sides.

You can use the square in testing this measure as shown in Fig. 5, from 25 to 26. Care must be used not to use too long a measure, by pressing the gauge into the arm. It is best to learn to use the tape.

Place the short part of the square upon the hip at the lower edge of the belt, with the long part running straight up, touching the back of the arm at the shoulder. Then slide up the gauge on the long part until it just touches the arm close to the body. Dont let the thin edge crowd up into the arms. The figures on the long arm of the square (on the side of the gauge which touches the arm), will give the correct measure.

FOURTH.—*The length of the back* is obtained by measuring from a point *one inch above* the prominent bone in the back of the neck, to the lower edge of the belt, or below the edge of the belt when you desire to extend the waist for hollow backs, as shown in Fig. 3, from 14 to 15.

CAUTION.—When the back is hollow and you wish to extend the waist at this point, continue the measure below the belt the desired length, and mark in the measure book the full length required, also mark opposite the next to the last question, the ¼, ½, or 1 inch, as the case may be, that the waist line at the centre of the back is lower than the hips. *When the belt goes about straight around* this is simply the distance below the belt you wish the waist line to be.

Fig. 2.

*When the belt is high in front and low in the back*, then you had better place the tape straight across from one hip to the other. Notice just how much the waist line in the centre of the back is lower than at the hips. This is the *amount* to be marked in next to the last question in the measure book. This extension of the back below the lower edge of the belt is provided for when setting the machine; full instructions for which are given in the instruction book, under the head of "Explanation for changes on the Machine," page

Another way for measuring the length of back, well adapted for use in the Drafting Machine, and especially for irregular forms, but seldom used, is as follows:

First, when measuring the width of back with the square, as shown in Fig. 4, with a piece of tailors' chalk, make a mark on the centre of the back just at the upper edge of the square. Then measure from the point of the neck down to this mark for the upper part of the back, and from the mark down to lower edge of belt for the lower part of the back, placing the figures for each part in the measure book, for which you will see separate spaces provided. The square should always be straight across the back, with the short piece close up under the arm and straight through, pointing neither up nor down, when you make the chalk mark.

FIFTH.—*The armhole* is measured at the shoulder, where the sleeve is to join the body. Have the hand resting on the hip and measure the arm rather tight, as the armhole gets larger while making up. The place to measure is shown on Fig. 1, at 27.

SIXTH.—*The length of shoulder* is measured from the neck to the arm hole, as shown on Fig. 4, from 28 to 29. It is measured just back of the shoulder on the shoulder seam. The machine allows for the seam. It is not necessary to take this measure unless the shoulder is very long or short, as we can get the correct length when we set the back of the machine. It is useful for a fleshy figure, but not necessary.

SEVENTH.—*The length of front:* Place the end of the tape line at the top of the breast-bone at the lower part of the neck, and measure down to the lower edge of the belt at the waist, as shown in Fig. 2, from 7 to 8. This will give a dress of medium height at the neck. If you desire it to be higher than medium you should place the end of the tape a little higher than the top of the breast-bone, and use a very tight neck measure.

*CAUTION.*—Never measure below the lower edge of the belt for the front. When you desire to extend the front below the belt, it must be done by changing the machine after it has first been properly set at the correct measure as taken.

Instructions for shortening or lengthening the front for peculiar figures are given under the head of "Explanations for Changing the Machine," page

EIGHTH.—*Height of front darts:* The height of the darts is obtained by measuring from a little below the point of the bust to the lower edge of the belt, as shown in Fig. 2, from 10 to 11. Measure for the *first dart only*, being careful not to start *too* high up. Darts should be measured too short rather than too long.

Fig. 3.

NINTH.—*The bust measure* is taken over the fullest part of the bust, not above it, close up under the arms and across the shoulder blades, keeping up a little on the back, as shown in front in Fig. 2, and at back in Fig. 1, at 4. It is best to take this measure while standing at the back, and always ascertain whether the same or a corset of a similar bust formation is to be worn with the new garments.

TENTH.—*The waist measure* is taken with the belt removed and medium tight, as shown in Fig. 1, at 5. Stand at the back while taking this measure. When taken over the belt, one inch must be deducted for the correct measure.

ELEVENTH.—*The hip measure* is taken *five inches* below the waist, around over the hips, rather loosely, as shown in Fig. 1 at 6. The quantity of clothing and drapery must be considered when taking this measure. It is best taken when standing at the back.

TWELFTH.—*Size of front darts* is obtained by subtracting the waist measure from the bust measure, thus: bust, 36, waist, 24; difference, 12, to go in the front darts. 12 is a medium size for front darts. When setting the machine, use two sizes less for princess or polonaise. For fuller information about darts see page

THIRTEENTH.—*The size of hip.* The hip dart for princess, polonaise and coats is graded as follows: Very small, 6; small, 8; medium, 9; large, 10; and very large, 12.

When using the sizes, you notice the hip, and place such a figure in the measure-book as indicates the size, or better still, measure with the square as follows:

TO MEASURE THE HIP DART WITH THE SQUARE, take the square in the left hand, with the short arm and gauge hanging down, and place the end of the long arm marked *A*, against the belt, having the lower edge of the square even with the lower edge of the belt. Then move the gauge up towards the body until its lower end or point just touches the hip slightly. The figures at the edge of the gauge nearest the body, on the *hip dart scale*, is the correct size. The position of the square when taking this measure is straight out from the body as shown in Fig. 4, at 23 and 24. Only used for princess, polonaise and coats.

FOURTEENTH—*The sleeve.* The length is taken with the arm raised to a horizontal position and bent at a right angle, measuring from the centre of the back to the elbow for one measure, and to the prominent bone of the wrist for the full length of the sleeve, as shown in Fig. 3, from 17 to 18 and 17 to 19. One-half the width of the back is to be deducted from each measure.

Fig. 4.

*For tight sleeves*, measure around the upper part of the arm half way between the elbow and the shoulder as shown in Fig. 3 at 20, and also measure around the arm at the elbow and over the hand and thumb. Have the fingers open and the thumb closed to the palm of the hand.

Sleeve length to elbow ....... .... ....

To wrist ........................ ....

Around the arm, *upper* part .. ....

Around at the Elbow....... ...

Around at the Hand............

FIFTEENTH.—*The length of skirt* is measured from the lower edge of the belt to the bottom of the skirt at front, side and back, as shown in Fig. 2, from 8 to 9 and Fig. 3, from 15 to 16.

Length of skirt:

Front..........................

Side .... ..........................

Back ............................

## QUESTIONS TO BE ANSWERED IN THE MEASURE BOOK.

At what point above the waist is the back the fullest? Is it ¼, ½ or ¾ the way up between the waist and the neck?

Answer..................................

Is the back very round or nearly straight?

Answer..................................

Is the back hollow at the point where the sleeve joins the body? and will it need to be fitted or padded at that point? At the top of circle of the back?

Answer..................................

Is the back hollow or lower than the waist at the hips? If so, how much, ¼ or ½ inch?

Answer..................................

Do you wish the back extended down, or do you wish to fill up to the belt with a bustle?

Answer..................................

Is the stomach high? That is does the belt slant up in front, instead of going straight around the body?

Is the stomach small, medium or large?

Answer..................................

If one shoulder or hip is higher or larger than the other, please notice it; also any other peculiarity of the figure.

Answer..................................

Fig. 5.

*When taking the measure for a garment, strive for exactness,* and always see whether the waist line is straight around, or if low on the back, how much.

When using the *Square* in taking measures, have the gauge placed upon the long arm so that it will be on the same side as the short arm of the square.

When using the square and gauge, hold the square in the *left hand*, and move the gauge with the right hand, placing the thumb partially upon the gauge and partially upon the square, so as to prevent its moving until you have read the measure from the square.

For marking all straight lines, remove the gauge and use the square.

## TO MAKE A LONG OR SHORT WAIST.

The best result is always attained when the lower edge of the belt is at the natural waist. If you wish to lengthen the waist, crowd the belt down, and this gives an increased underarm length. If you wish to shorten the waist, place the lower edge of the belt at the point where you wish the waist to terminate. Or, in other words, for a long waist use a long underarm measure, and for a short waist use a short underarm measure.

## RESULTS OF BAD MEASURING AND SUGGESTIONS FOR REMEDYING THE DEFECTS ARISING THEREFROM.

REMEMBER THAT AN OUNCE OF PREVENTION IS WORTH A POUND OF CURE, THEREFORE BE CAREFUL TO MEASURE CORRECTLY.

*When the bust measure is too large* there will be a fullness under the arms or in the back. Take up at the side seam.

*When the bust measure is too tight*, add goods at the side seam. If the buttonholes are not made add some goods down the front line, and sometimes in the centre seam of the back.

*When the back measure is too wide*, take up the centre seam of the back and add to the front down the front line, or if the buttonholes are made you might trim out the armholes at the back a trifle, and increase the size of the sleeve at the armhole accordingly.

*When the back measure is too narrow*, the front will be wide and there will be fullness in front of the arms. Give what goods you can in the centre seam of the back. Take up the front on the fold line, and trim the armholes in front.

*When the armhole measure is too large* there will be fullness back of the arm. Take up at the shoulder seam back of the arm.

*When the armhole measure is too tight* it will bind over the arm at the shoulder, and draw the goods up from the hip, making it short at this point. Let out the shoulder seam back of the arm.

*When the length of back measure is too long* the garment will be too high back of the neck, and wrinkle across the back. Take up the shoulder seams and trim the neck.

See page 13 when underarm measure is too short.

*When the length of back measure is too short* the garment will bind each side of the neck and be short at the waist on the back. Loosen the shoulder seams and drop the back down to its place. This will make it low at the back of the neck. Piece the lining, and if the goods are cut, hide the piecing by trimming, or with fancy collar.

*When the front measure is too long* it will be too high in front at the neck, and too loose on the shoulders at the neck. Take up the shoulder seams and cut out the neck and front.

See when the underarm length is too short.

*When the neck measure is too large* it also will cause a looseness on the shoulder near the neck. Take up the shoulder seam.

*When the front measure is too short* the garment will be low in front of the neck and be tight or bind on either side thereof. Loosen the shoulder seams.

*When the neck measure is too tight* it also will bind back of the neck, but will not be low in front. Loosen the shoulder seams.

*When the underarm measure is too long* the garment will be too low at the neck, tight on the shoulders near the neck and extended down at the waist so as to cause wrinkles. Loosen the shoulder seams and draw the garment up to its place. Fit on the shoulders, and trim out the armhole under the arm.

*When the underarm measure is too short* the garment will be too high at the neck and short waisted. Loosen the shoulder seams, and drop down until right at the waist.

*When the waist measure is too large* the back will be loose. Take up the side seam.

*When the waist measure is too tight* add goods at the side seam.

*Don't mistake the hip dart or underarm seam for the side seam in making alterations.*

*When the hip measure is too tight* the skirt will ride up and wrinkle at the waist. Add goods at the side seam, and if more are required, at the fold line in front, and at the centre seam of the back.

*If the hip dart is too small* it also will cause the skirt to ride up and wrinkle at the waist. Add goods at the side seam, or take up the hip dart at waist if possible.

*Three things cause wrinkles at the waist*, the underarm length being too long, the hip dart being too small, or the skirt measure over the hips too tight. Ascertain which of these causes the difficulty, and remedy as suggested. The safest course while learning, is to take a loose hip measure.

Bad basting may cause wrinkles anywhere in a garment. In fact the most perfect garment can be so distorted by putting it together that it will be spoiled. Learn to baste. See HINTS TO DRESSMAKERS.

## SAMPLE MEASURE.

THIS MEASURE IS USED FOR PRACTICE THROUGHOUT THE BOOK

| | Inches. |
|---|---|
| Neck | 13 |
| Width of back | 13 |
| Length of back | $16\frac{1}{4}$ |
| Underarm Length | 8 |
| Armhole | 15 |
| Shoulder Length | $5\frac{3}{4}$ |
| Front Length | $13\frac{1}{2}$ |
| Front dart height | 5 |
| Bust measure | 37 |
| Waist measure | 24 |
| Hip measure | 41 |
| Size of front darts | 12 |
| Size of hip | 9 |
| Length of skirt, front | 38 |
| " " side | $38\frac{1}{2}$ |
| " " back | 39 |

Sleeve length to elbow, $18\frac{1}{2}$; to wrist, $27\frac{1}{2}$; less one-half width of back, $6\frac{1}{2}$ inches; leaves length of sleeve to elbow, 12; and to wrist, 21.

Sleeve width at upper arm, 12; at elbow, $11\frac{1}{2}$; at wrist, 8.

Always answer the following questions in your measure book, as certain changes in drafting depend on them.

Is the back round or hollow? If so, how far above the waist? ........

Do the shoulder points drop to the front? That is, must you fit back of the arm, where the sleeve joins the back, or at the top of the side seam?..

Is the waist-line in the centre of the back lower than at the hips, and if so, how much? $\frac{1}{4}$, or $\frac{1}{2}$ inch?........................................

Is the stomach high? If so, how much? $\frac{1}{4}$, or $\frac{1}{2}$ inch?................

## TO USE THE DRAFTING MACHINE.

Apply the several measures taken each to the part arranged for it. On the machine near each scale you will find figures in [ ] which tell you which part of the machine is to be fixed first, second and so on. Near the figure in the brackets [ ] you will find the name of the measure you are to get from the measure book and use for that particular scale.

Thus at the armhole part of the back you will find "[1st]," and under it the word "Armhole." This means that you are to set this part first, according to the armhole measure. Next across the back you will find "[2d] width of back," which means that this is the second part to fix, and is to be set at the width of back, and so on.

ALWAYS SET THE MACHINE IN THE FOLLOWING ORDER:

*First, set the* BACK-PIECE.

*Second, set the* SIDE-PIECE.

*Third, set the* UNDERARM-PIECE.

*Fourth, set the* FRONT-PIECE.

The figures on cuts No. 6, 8, 10 and 12, represent the order in which the machine is to be arranged. Looking at cut 6, the Back, we find (1) at the armhole; this indicates that this is the first part of the back to be set. We find in our measure for practice that the armhole is (15), so we set this part of the machine at (15).

Next set the part marked (2) which is the width of back, to the measure given which is (13), and then set the parts marked 3, 4, 5, 6, 7, and 8, in order named to their several measures, as given.

In the following instructions, the figures on the *left* side correspond with the figures on the cut, and those in brackets [ ] stamped on the machine, represent the order of setting the different parts of the machine, and figures on the right of the instructions are taken from the sample measure and these measures are to be found in the scales on the machine.

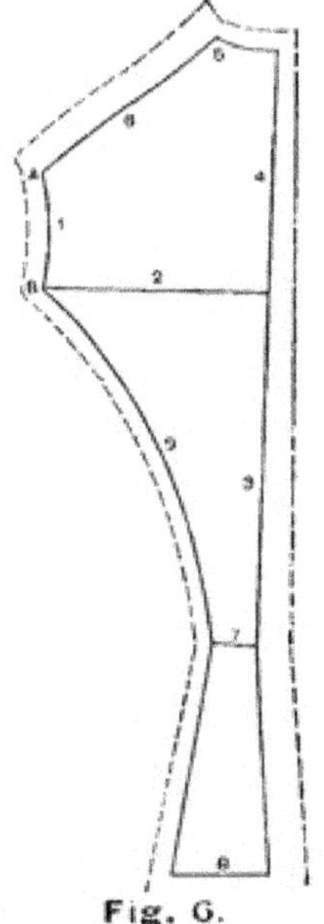

Fig. 6.

## THE BACK.

| | | SAMPLE MEASURE. |
|---|---|---|
| 1st. | Set 1, (see Fig. 6), the armhole to the measure........................ | (15) |
| | You will find two places to fix the arm-hole on the back, the upper shown by A, in Fig. 6, and the lower by B. Both are set at the armhole size, in this case 15, put the centre of the screw on the number. Have this part of the armhole, A B, straight up or parallel with 4, the centre line of the back for a medium length of shoulder. | |
| 2nd. | Set 2, the width of back, to the measure | (13) |
| 3rd. | Set 3, the lower part of length of back, to the underarm length .......... | (8) |
| 4th. | Set 4, the upper part of length of back by simply adding enough on the upper part to what you have used on the lower part to complete the back length. Thus, length of back 16¼; used for lower part of back length 8, add for upper part 8¼ to make the 16¼ for full length...... | (8¼) |
| 5th. | Set 5, the neck, to neck measure..... | (13) |
| 6th. | Set 6, the shoulder, so as to bring the point of the armhole marked A on the cut straight up over the point marked B, for medium style. You can make the shoulder long or short by moving point A, Fig. 6, to the right or left...................... | (3[illegible]) |

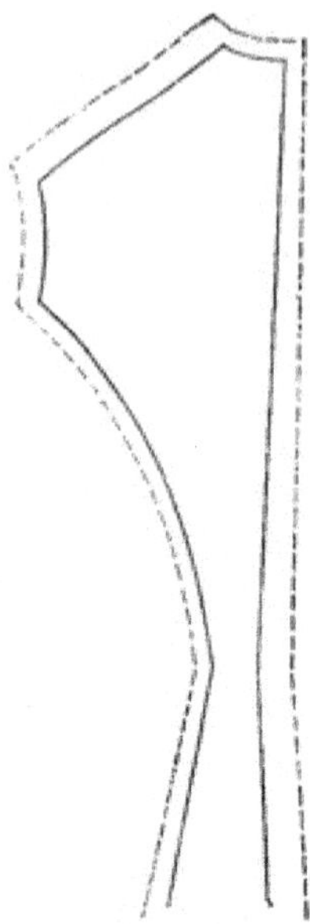

Fig. 7.

7th. Set 7, the back of the waist, according to fashion, wide or narrow. Place the centre of screw at size 3 for medium or a 24 waist .... ...... (3)

8th. Set 8, the skirt of back by placing the centre of screw at the same figure that you used at the waist above. That is, if you set at 3 at the waist, also set at 3 for the skirt.......... (3)

*To Mark the Back.*—Mark outside for cutting, and inside for the sewing lines, and at the lower edge of cross-piece at the waist for the waist line. To get the hollow of the back, mark a line from the wide seam line at waist line, up to the fullest part of the back, generally ⅔ of the distance from the waist to the neck, opposite B, Fig. 6, then gradually deepening the seam from that point down to the waist. This will give us Fig. 7.

CAUTION.—Number 3, or the lower part of the back length, is always set at the underarm length where the waist line is straight around. But when the waist line is lower in the centre of the back than it is at the hips, then we must add to the underarm length for No. 3 on the machine just as much as the waist line drops in the centre of the back. Thus, if the waist line drops ½ an inch on the back, then we add the ½ inch to the underarm length. In this case the underarm length is 8, to which we add the ½ inch making 8½ for the lower length of back. If the back is 16¼ long and the lower part 8½ the upper part is 7¾, this makes the back length correct.

It is very important to learn just how much the waist line is lower in the centre of the back than at the hips, as it has to do with the fit on the shoulder.

## THE SIDE BODY.

*That part of the side body which joins the back is set first.*

1st. Set 1. (See Fig. 8), the circle of side body, at the same figures as you find on the circle of the back designated by figure 9, on cut 6, diagram of back, in this case, it is .... (11¼)

2d. Set 2, the armhole part by placing the centre of the screw on the armhole size......... (15)

3d. Set 3, the side seam, by the underarm length (8)

4th. Set 4, the waist, at the waist measure, less the number of sizes used on the back at the waist. Thus: waist 24, less 3 used in the back, leaves 21, which is the number to use here........................... (21)

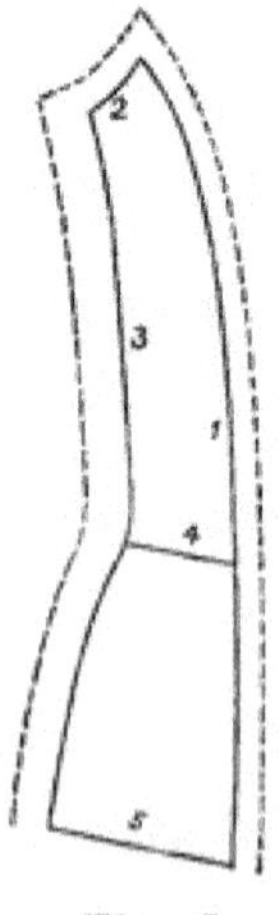

Fig. 8.

Remember the number of fashion sizes used on the back at the waist, must come off the side body at the waist.

5th. Set 5, the skirt, at the hip measure, less the number of sizes used in the skirt on the back. Thus: hip measure 41, less 3, the number used on the back, leaves 38.... (38)

*To Mark the Side Body.*—Mark outside for cutting, and inside for sewing lines, and at the lower side of cross-piece at waist for the waistline. We then get Fig. 9.

**Fig. 9.**

CAUTION.—Number 3, Fig. 8, is set at the underarm measure except when you extend No. 3 on Fig. 6. When you add ½ inch to the lower part of back length, add half as much to No. 3 on the side body. That is, the circles of the side body which joins the back are always set at the same figures as the circles of the back, but the part of the side body which joins the underarm piece is set at the underarm length, except when you add to the lower part of the back length on account of hollow back, then add only half as much to No. 3; this part of the side body.

## THE UNDERARM PIECE.

SAMPLE MEASURE.

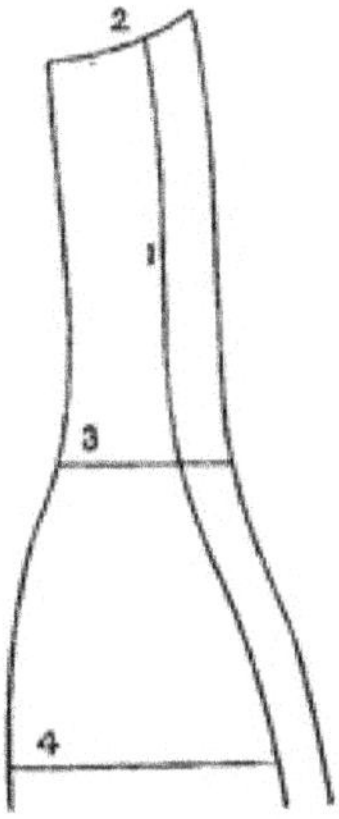

Set the part of the underarm piece which joins the side body first. See Fig 10.

1st. Set 1, the side seam to the underarm measure, in this case.............. (8)

2nd. Set 2, the width at the top or armhole, at the bust measure.............. (37)

3rd. Set 3, at the waist measure ........ (24)

4th. Set 4, the skirt at the hip measure ... (41)

Remember, you are not to take off any sizes at the waist and skirt on the underarm piece, as you did on the side body. That is only done on the side body to offset the changes on the back to follow fashion.

**Fig. 10.**

*To Mark the Underarm Piece.*—Mark the pieces running up and down outside and inside, across the top and across on the lower edge at the waist line. The outside lines are the cutting lines, the inside sewing lines, as shown in Fig. 11.

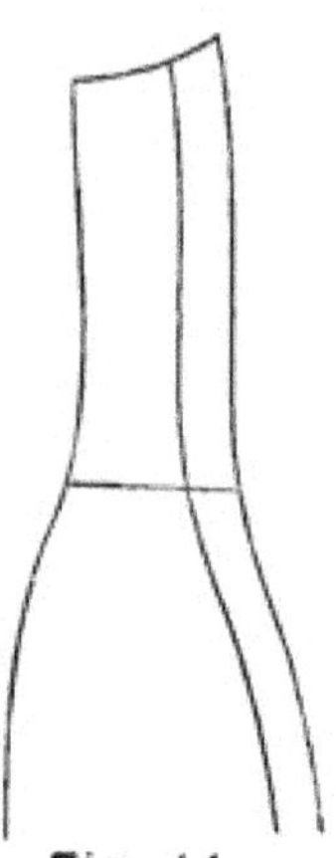
**Fig. 11.**

CAUTION.—In setting this part of the machine you will find that whenever the waist measure comes within 8 inches or less of being as large as the bust measure, that the piece will be much larger at the waist than will look well. This you can correct by taking off 3 sizes at the waist at No. 3 and adding the same amount on the front at the waist at No. 14, Fig. 12. Whenever this piece gets wider or narrower at the waist than at the top, it only follows the shape of the customer and gives just the shape required to fit. So when we add to the front at the waist what we take off here at the waist, we keep the same amount of goods.

## THE FRONT.

SAMPLE MEASURE.

*Always set that part of the front which joins the underarm piece first.* See Fig. 12.

1st. Set 1, the side seam, at the underarm length.............. (8)

2nd. Set 2, the length of front, lower part, at underarm length . (8)

3rd. Set 3, the length of front upper part, by adding enough to the amount used in the lower part to complete the length of front. Thus: length of front 13½, used for the lower part 8, add 5½ for upper to make full length........... (5½)

4th. Set 4, the neck, at measure.......................... (13)

5th. Set 5, the shoulder, to same figure as indicated on the shoulder of the back.......................... (5¾)

6th. Set 6, the armhole, upper part, at the measure ........... (15)

7th. Is always set.

8th. Set 8, the width of front, to the bust measure, less width of back. Thus: bust measure 37, less 13, width of back, leaves 24 for width of front....... (24)

This is generally set 1 size less for small bust, 2 for medium, and 3 for very large, to make the curved front.

9th. Set 9, top of first dart, at same figure as width of front above. When width of front is 24 put top of 1st dart at 24. This figure gives you a medium width, but you can vary the width according to taste (24)

10th. Set 10, top of second dart, at the same figure as width of front. When front width is 24 put top of second dart at........................... (24)

11th. Set 11, height of dart, at medium, or according to measure.......................... (5)

12th. Set 12, size of front darts at waist, at 12 for medium, or at size in measure book ...................... (12)

See how to get the size of darts on page

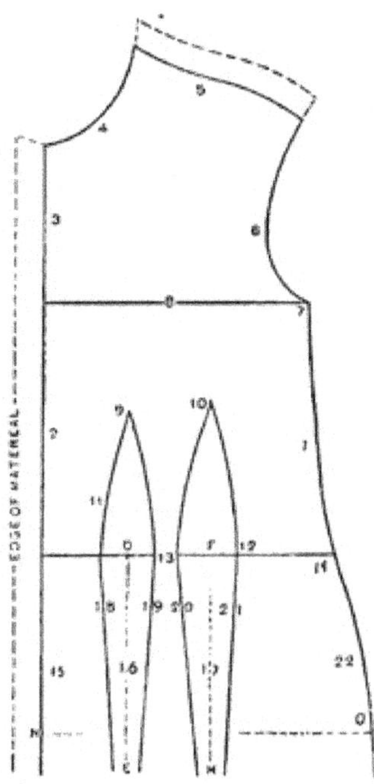


**Fig. 12.**

13th. Set 13, so as to make the second dart a little larger than the first, between the figures 4 and 5 for medium, unless otherwise desired . . . . . . . . . . . . . . . . (4 to 5)

14th. Set 14, the waist size, at the waist measure, fasten the centre of screw on the size. . . . . . . . . . . . . . . . . (24)

## TO DRAFT THE GARMENT BEGIN WITH THE FRONT.

### To Mark the Front.

Have the front edge of the machine back from the edge of the paper or lining 1½ inches, and the waist line the distance from the bottom that is desired for the length of the skirt. Then begin at the waist line and draw a line up along the edge of the machine to the neck. This gives the fold line.

Then follow the edge and mark around the outside for the *neck, shoulder, armhole* and *underarm seam*, down to the waist line. Then mark the inner edge of the shoulder and the inner edge of the underarm seam for sewing lines. Then mark from underarm seam along the lower edge of the machine to the front, for the waist line.

Then mark the dart nearest the front, so as to bring it to a point at the top. Then mark the second dart same as the first. Mark on the inside edge only of the dart pieces for sewing lines.

*These dart marks are the sewing lines. The cutting lines are the straight lines at D and F, up through the centre.* See Fig. 12.

## THE FRONT SKIRT.

To add the skirt or continue the front below the waist line: First, extend the fold or front line, 15, straight down to the bottom of the basque; then draw line 16 from D, at centre of first dart at waist line straight down or with the grain of the lining to the bottom of the basque at E. Mark lines 18 and 19 so as to extend the dart *not from above the waist line but* from the waist line down to the bottom of basque, so as to have the lines 18 and 19 one-third of an inch from line 16 at E. This is allowed for the seam. Mark the second dart same as first, by first making line 17 from F in the centre of the second dart straight down to H. Then mark lines 20 and 21, as shown on the diagram and figure 12.

To get the proper spring for line 22, find the width of the front skirt as given in the *skirt table for front*, opposite the hip measure on page ; measure on the hip line (*which is five inches below the waist line*) from N, on line 15, to O, on line 22, the length given in table, *always omitting* the spaces occupied by the darts. Measure this width with the tape, begining at N, on line 15, measure to line 18, then move the tape along to line 19, and measure to line 20, again move over to 21 and continue the measure to line 22, and complete this part of the hip measure for the front, which for 41 hip measure is 9¼. This gives the correct position for point O. If the skirt table is not handy, take and divide the hip measure by 4, and then take off 1 inch, the remainder is the width for the front skirt without darts or seams.

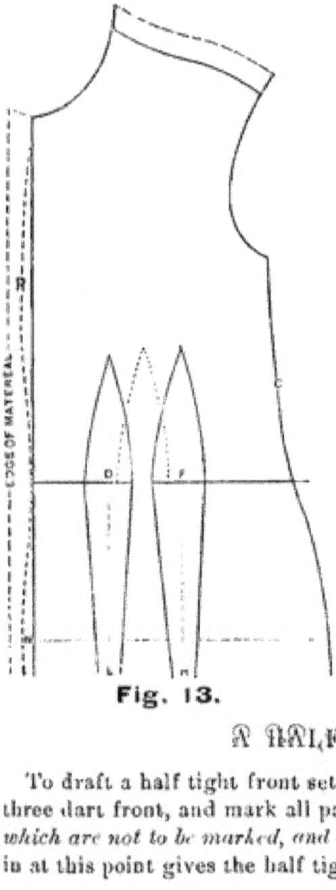

Then, using the outer edge of the wide piece of the side body, place the point of the wide piece of side body at the end of waist line, at 14, Fig. 12, with the outer edge passing through point at O, and mark along this edge from the waist line to bottom of basque, and this will give line 22. Then move to the right ⅓ inch and mark along the same edge for cutting line.

The dotted line R, in Fig. 13, shows the curve or swell front.

Fig. 13.

## A HALF TIGHT FRONT.

To draft a half tight front set the machine the same as for the regular three dart front, and mark all parts the same, except the *two front darts, which are not to be marked, and are not to be taken out.* Leaving the goods in at this point gives the half tight front.

You need not add anything to the waist size as you get looseness enough by not taking out the two front darts.

Loose sacques and wrappers can be made on this plan.

## A THREE-QUARTER TIGHT FRONT.

To make a three-quarter tight front set the machine the same as for the regular two dart front, as instructed on page 15, and mark the same as there instructed, except the two front darts. Then move the machine to the left so that the second dart is half way between the hip dart and the fold line, or between where the first and second darts would be if marked. Then mark the second dart. The position of this dart is shown by the dotted line on Fig. 13. This dart is extended below the waist as usual.

### Skirt Rule for Front of Basques, &c., 5 Inches below the Waist Line.

| FIND IN THIS COLUMN THE HIP MEASURE. | WIDTH OF FRONT WHEN FINISHED. SEAMS AND DARTS NOT INCLUDED. | FIND IN THIS COLUMN THE HIP MEASURE. | WIDTH OF FRONT WHEN FINISHED. SEAMS AND DARTS NOT INCLUDED. |
|---|---|---|---|
| 35 | 7¾ | 49 | 11¼ |
| 36 | 8 | 50 | 11½ |
| 37 | 8¼ | 51 | 11¾ |
| 38 | 8½ | 52 | 12 |
| 39 | 8¾ | 53 | 12¼ |
| 40 | 9 | 54 | 12½ |
| 41 | 9¼ | 55 | 12¾ |
| 42 | 9½ | 56 | 13 |
| 43 | 9¾ | 57 | 13¼ |
| 44 | 10 | 58 | 13½ |
| 45 | 10¼ | 59 | 13¾ |
| 46 | 10½ | 60 | 14 |
| 47 | 10¾ | 62 | 14½ |
| 48 | 11 | 64 | 15 |

## EXPLANATIONS FOR CHANGING THE MACHINE.

***To Extend the Back Below the Belt for Hollow Backs.***

The lower part of length of back, scale 3, in Fig. 6, is extended down one-quarter or one-half inch, according to the memoranda made in the measure book at the time of taking the measure in answer to the question, "Is the back hollow below the belt on the hip?" Just as much as you have decided that the waist line in the centre of the back is lower than at the hips, just that much you are to add to the underarm length for lower part of back length on scale, 3 Fig. 6, and as much as you add on scale 3 you take off scale 4 for the upper back length, so as to keep the back length correct. Don't forget when the belt is low on the back, or the waist line slopes down, to notice how much lower it is than at the hips, and make the amount ¼, ½, 1 inch, as it may be, for this is needed to keep the correct slope of shoulder.

***To Lengthen the Waist in Front.***

After the machine is all set extend the lower plate of the front (scale 2, Fig. 12) the amount desired, say one-half inch. This is used in crowding the waist down in front.

***To Shorten the Waist in Front for a High Stomach.***

Take off one-quarter or one-half inch from scale 2, at bottom of machine in front Fig. 12, and add the same amount to scale 3, at the top, thus preserving the correct length of front.

***To Change the Slope of the Shoulder in the Front.***

When the length of back is very short and the length of front very long, the front shoulder will slant on the machine very much. To correct this you can add ½ an inch or more to the lower scale on the front length, and take it off the upper. This will keep the length of the front the same and bring the shoulder down some at the neck. When you make this change don't forget to set the height of dart to the measure.

***To Raise the Shoulder Seam at the Arm-hole.***

To give a square shouldered back, add ¼ or ½ inch to the armhole of the back, and take off the front armhole the same amount.

***When the Shoulder Blade is Prominent, and there is a Hollow Place Back of the arm that Needs Fitting.***

Draft the side body as usual, stand the pencil along the circles 3 inches from the armhole, then with the pencil as a pivot move the armhole part down ¼ inch. This makes the armhole part that much narrower and lengthens the circles ¼ inch so they will be the same length as the back circles, when this is required you so mark, in the measure book.

## DARTS.

The Drafting Machine enables us to use all systems of darts. You can use the old rule of taking out in the front darts the difference between the waist and bust measures. Thus, bust 37, waist 24, difference 13, to come out in front darts. As we always use a curve front, we use 1 size less, making the front darts 12. To get rid of the fullness at the bottom of the front darts in princess and polonaise, we use 2 sizes less in the front darts. The rule just given is generally correct, but when the bust is large and the stomach small, this rule gives too much goods below the waist line in princess and polonaise, and when the stomach is large and the bust small, it fails to give enough goods over the stomach. Now as our fullness for the bust does not depend entirely on the darts, we vary them to suit the stomach. When the stomach is larger than the bust, in proportion we add a size or two to the darts, and when the stomach is very small and the bust large, we take off one or two sizes in the darts, this enables us to have the proper fullness at the bottom of the front darts in polonaise and princess.

When the waist and stomach are large and both project nearly the same, small darts are needed, when the waist is small and the stomach large, there is more spring out from the waist, and large darts are needed to give the proper amount of goods. When the stomach is straight down, use 10. If it projects out from the waist the usual amount use 12, which is medium. When more than medium 13. When very large, projecting very far 14.

The tops of the front darts can be arranged to suit the prevailing style, or the individual taste of the customer. Generally they are set at the same figures as width of front, found by taking the width of the back from the bust measure. Thus, 37 bust, 13 width of back, leaves 24 for width of front ; set tops of 1st and 2d darts at 24. They can, however, be drafted in any position, as is explained in detail in a following paragraph.

***How to Draft the Darts in Any Position.***

There are styles and figures which require the position of the darts to be changed from those given on the machine. For example:

When the waist is over 28 inches, the first dart would be too near the front line, and the space between the first and second darts too small. To remedy this, set the machine regularly, as instructed on page 15, and draft as before, all except the darts. Then, if the front dart is too near the front line, move the machine to the right from ¼ to ½ an inch, as desired. Mark the first dart, then move the machine ¼ inch more to the right and mark the second dart. This places the first dart farther from the edge and the darts farther apart. By moving the machine you can make the darts higher or lower than the machine gives them, when you so desire.

### *How to Leave the Hip Seam Out.*

Set the machine as directed on pages 11 and 15, for the regular basque mark the front, then place the underarm or hip seam of the underarm piece directly over the underarm or hip seam of the front, and mark the armhole and side seam. This style of basque generally stops at the waist and can be used with a belt.

### *How to Change the Underarm Seam into a Hip Dart.*

Set the machine as directed on pages 11 and 15, for the regular basque mark the front and add a skirt six inches long, as directed on page 16, then place the under arm-piece about 3 inches to the right of the front so its waist line is on a line with the waist line of the front, that is, have about 3 inch space between the underarm seam of the front and the underarm seam of the underarm piece. Mark the underarm-piece when in this position above the waist line and then below for skirt 6 or 7 inches. Mark both cutting and sewing lines. Have the sewing lines of the underarm or hip seam of both the front and underarm piece come together 6 or 7 inches below the waist line.

Now the exact distance to have the underarm piece from the front is controlled by the size of the hip as found by the square. If it is nine sizes then have the under arm-piece just 9 sizes to the right of the front.

### *When the Bust Measure is Too Large for the Machine,*

OR, HOW TO DRAFT THE FRONT WHEN BUST MEASURE, LESS THE WIDTH OF BACK, EXCEEDS 33 INCHES, WHICH IS THE LIMIT OF SCALE 8, FIG. 12, ACROSS THE FRONT.

If the bust measure is 51 and the width of back 16, this would leave for width of front 35 inches. We open the front of the machine to 33 the limit of its scale, and this will leave two sizes less than the amount desired. These two sizes, equal to ⅔ inch (a size on this scale being ⅓ inch), we add to the left of the fold line in front, opposite a point 1 inch above the top of the front dart, at the fullest part of the bust. This changes the fold line from a straight to a curved line, as shown by the dotted line R on Fig. 13.

In case you should ever have to draft a front exceeding 35 sizes in width, then allow goods under the arm and at the side seams and a full inch extra down the front. This is obtained by moving the fold line of the machine back to the right from the edge of the goods.

### *When and How to Use a Curved Front.*

*A curved front is used for most figures*, and is necessary on all those with large busts, and wherever there is a tendency for the goods to cross or lap in front at the neck.

If you wish to use a curved front when setting the machine, take off from the measure for the *width or front* 1 size for small curve, thus, with the bust measure 37, width of back 13, leaves 24 for the width of front.

Instead of setting the machine at 24, set it at 23, thus allowing one size to be added beyond the fold line for the curve. *This is added at a point opposite the fullest part of the bust, which is one inch above the top of the first dart.* To draft this curve, draw a line *from this point* to the fold line *at the neck* and to the fold line *at the waist.* Curve these lines slightly, using the curve of the side body, or with the tracing wheel on the lining.

Use 1 size for a small curve, 2 sizes for medium, and 3 sizes for very large.

Whenever the front line curves more than one size, take a ½ inch plait in the lining at R, Fig. 13, and extend it two inches to a point towards the top of first dart. This will give a broad look across the bust much like a corset.

### *How to Make the Bias Cut Under the Arm.*

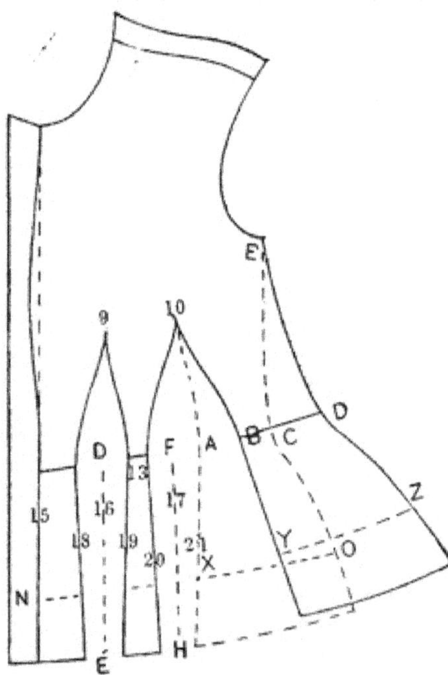


Fig. 14.

Set the machine as directed on pages 11 and 15. Add 3 or 4 sizes to the second dart by moving A to the right to B as shown by Fig. 14, this moves C to the right to D, and makes the goods from the second dart to the under-arm seam more or less bias as desired.

If the bust is 36 and the waist 24 the rule gives us 12 difference to go into the front darts. Now to make the bias under arm we add 3 or 4 sizes to the 12 and put 15 in the darts.

The skirt is added by extending line 15 straight down. Line 16 is also straight down from D to E. Lines 18 and 19 extend from the waist to ½ of an inch each side of E. Line 17 is extended from F to H straight down, the same as on page 15, Fig. 12 and line 20 and dotted line 21 from the waist to ½ of an inch to each side of H. Line C, O, gets its spring from the front skirt table which gives distance from N to O, darts excepted ; this gives the skirt for a regular basque. Now to get the bias skirt we increase the second dart from A to B, then we move X to the right to Y, so the line Y B comes to the dart line B 10 at the waist in a straight line or nearly so. This carries O over to Z, as the length from X to O is the same as from Y to Z extend line E D on through Z with the regular curve. The amount of goods from N to Z is the same as from N to O. The bias cut under the arm is an old idea revived. The advantage in having the cloth on the bias is that the goods draw in nicely over the hip and above it reaching to point E at the arm-hole.

The disadvantage is in the awkward look in striped goods, and in the fullness at the top of the second dart, at 10. With our plan of getting the size of the hip you ought to fit nicely without the bias under arm. As you can and must do this in Princess and Polanaise you can do it in basques as well. But some times the customer's mind is to be fitted or suited so follow fashion.

### *How to Make a Double Breasted Front.*

After having arranged the machine for the regular front according to the required measure, as instructed on page 15 or on the diagram, simply place it on the lining, 2½ or 3 inch back from the edge, to allow more goods for lap.

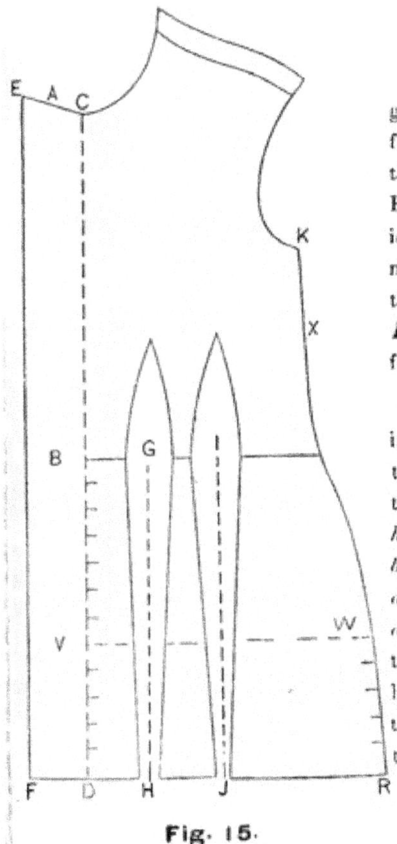


Fig. 15.

Three inches back from the edge gives a wide lap. Place it nearer or further from the edge according to the width of the lap desired. In Fig. 15 on this page, the centre line is the dotted line ***C D***, made by marking in the slot from the neck to the waist. The regular fold line ***A B*** and the line ***E F*** shows the width for double lap.

Extend the skirt as before, carrying the darts to the bottom so that there will be about ¾ inch of goods there for seam. *Remember that the hip measure is always applied 5 inches below the waist line, as shown by dotted line* ***V W***, Fig. 15, *and not at the bottom of the basque.* In extending the skirt below the 5 inches have the underarm seam spring to the right ¼ inch for each inch below the five.

### *To Make a Tight Front With One Dart.*

Set the machine for the regular front, as directed on page 15. Use curved front. Move the top and bottom of the first dart as far to the left as they will go and fasten the screws there, thus shutting out the front or first dart. Then move the right hand piece of the second dart three sizes to the left at the waist ; thus, if you had marked the front darts 12, after the change they would stand at 9.

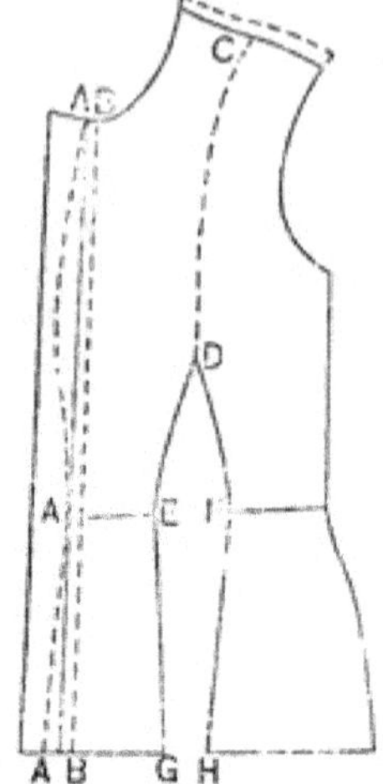

Fig. 16.

Move the top of second dart to the left, to about 3½, or so as to give the dart a good shape, as shown in Fig. 16.

Place the machine properly on the goods, and then mark the front edge, neck, shoulder, arm-hole, under arm, seam and waist line. Mark the dart. Curve the front as shown by dotted line *A*, on Fig. 16. The skirt in the front is the same as in Fig. 12, or the regular front, excepting there is but one dart in the front instead of two. The same rules apply to drafting this skirt as to the others. Remember that the hip measure in drafting is always applied *five inches* below the waist line and not at the bottom of the basque.

The regular back side body and underarm piece as described on page 12 and 13 can be used with this front.

### *How to Make a Seam From the Top of a Dart to the Shoulder or How to Make a Vest Front.*

This is done by extending the dart, as is shown by dotted line *C D*, Fig. 16. The line continues straight up from the top of the dart for about three inches, then curves slightly to the right, and then straight to the shoulder. It should reach the shoulder near the centre. Sometimes, however, it is carried nearer to the neck. Fashion regulates this. When the front has two darts, the first dart is the one usually extended.

When you make a seam from the top of the dart to the shoulder, you had better make a paper pattern and cut it apart on the line from the dart up, so you can allow for the seams in the goods. Do the same when you make a vest front.

### *How to Make a Basque with one Dart, Open on the Back.*

Set the machine the same as for the regular two dart front directed on page 15, or in measure book. After the machine is all set, shove the first dart, top and bottom, over to the left as far as it will go, thus shutting it out. Then move the right-hand piece of the second dart three sizes to the left. Thus, if you had set the front darts at 10, after the change it would stand as 7. Place the top of the second dart so it will be over the centre.

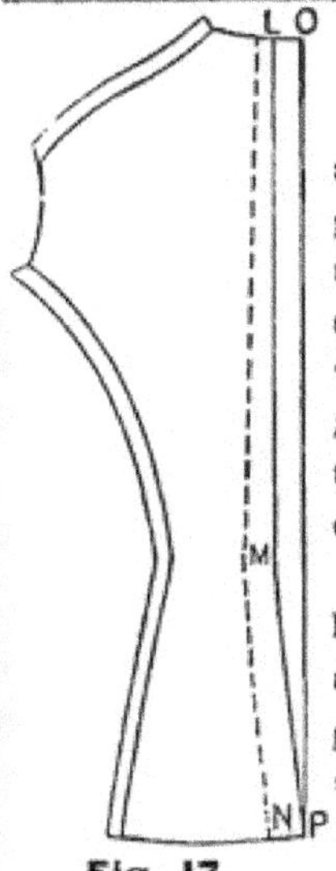

Fig. 17.

(*Do not use a curved front with this style.*)

Fold the goods and place the machine upon them, and instead of placing its edge at the edge of the goods as heretofore, move the machine to the left until you can see the edge of the goods through the slot to the right of the edge, this is the narrow slot in which the screws are placed, *and is the centre line of the front*, shown by ***B B***, Fig 16. When the machine is thus placed on the goods, mark as before, and also extend the skirt as usual.

We have you fold the goods, because you desire to have no seam in front. The object of moving the machine the left before marking, is to get rid of the goods that the machine allows for fold, and lap in the usual style of open front.

## BACKS OF VARIOUS KINDS.

### *Back Open up the Centre, to go with the Front just Described.*

The back side body and underarm piece to go with this front are set and drafted the same as usual, except that the cutting line down the centre of the back, shown by ***L M***, Fig. 17, becomes the fold line, and goods must be allowed to the right of this line for fold or lap—say about 1¼ inches—as shown by line ***O P***, Fig. 17. This completes a garment open at the back.

### *How to make a Sacque Back.*

Arrange the back and side body as usual, as shown on page 11, and in measure book, and mark both the cutting and sewing lines of the back except the circles where it joins the side body. Then dot the *sewing line* of the circle, two inches from the arm-hole towards the waist, and dot the *sewing line* at the waist. Next place the side body so that its *sewing line* will be over the dotted sewing line of the back, at ***E***, the arm-hole, and at ***C***, the waist, as shown in fig. 18, taking care to have the arm-hole at ***E*** and the waist line at ***C*** in place. Mark the arm-hole, the side seam and the skirt, as shown in Fig. 18. This back is simply the regular back ***A*** and side body ***B***, as made by the machine, drafted in one piece.

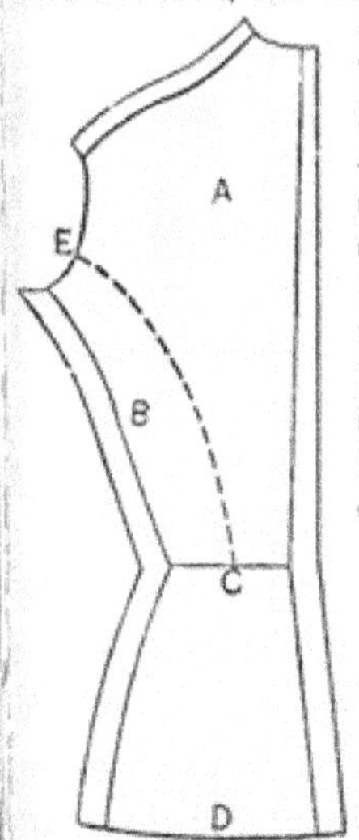

Fig. 18.

### *How to Make a Back With one Seam to the Shoulder.*

Draft the back and side body together as directed in the preceding directions for a sacque back. Place a dot on the waist line 1½ inch to the left of the centre line, or at ***C***, as shown in Fig. 19. Divide the back just below the shoulder into two equal parts at ***F*** from the seam line at the arm-hole to the

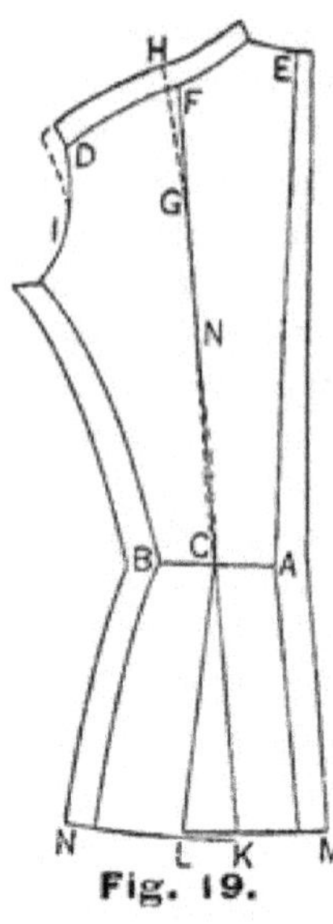

Fig. 19.

seam line at the centre of the back, just at the neck, as shows from *D* at arm-hole to *E* at neck in Fig. 19. Then draw a line from *F* at the shoulder to *C* at the waist. Extend this line straight on to the bottom of the skirt at *K*, for the side body.

To get the line *C L* for the skirt of the back, place the back of the machine in the same position it occupied when you mark the back, and move it to the left or right straight with the waist line until either the sewing or cutting line comes to the point *C*, then mark down that edge, and it will give you the line *C L*, as shown in Fig. 19.

Next mark the dotted line *H G* at the shoulder, that is curve the side, back to the left about ½ inch. In order to keep the shoulder at its proper length, add a piece at the arm-hole, as shown by the dotted line *I*, equal to the piece cut out in making the curve at *H G*. This gives a better fit back of the arm.

For this back it is best to draft a paper pattern, and cut the back apart on the line *L C F*. The skirt of the side body will require a piece to be added the size of *L C K* to give its proper width. Lay these pieces on the lining and allow ½ inch for seam where they cut apart.

The line where the paper was cut is the sewing line. In dividing the back at the waist line, the two pieces may be equal, but the side body must never be narrower than the back.

### *How to Make a Back with six Pieces with Seams to the Shoulder.*

Draft the back and side body together as for a sacque back, as shown in Fig 18. Divide the waist line into *three* equal parts, and also divide the distance from the sewing line of arm-hole at the shoulder to the sewing line at the centre of the back at the neck, into *three* equal parts. Draw straight lines from the divisions at shoulder to the divisions at waist, and extend the skirt as shown in Fig. 19. It will be necessary to draft this on paper and allow one half inch for seams where the pattern is cut in two.

### *How to Make a Back in six Pieces with one Straight line to the Shoulder and a Curve to the Arm-hole.*

First draft the back and side-body together, as shown in Fig. 18. Divide the back at the waist line into *three* equal parts and at the shoulder into 2 equal parts from the seam at arm-hole to the seam at centre of the back at the neck. Draw a line from dot at shoulder to dot at waist line nearest the centre of the back. Extend the skirt below the waist, as shown in Fig. 19. Next, with the circle of the side body, mark a line from that point in the arm-hole where the seam generally goes one-third at the distance from the top of the side-seam to the shoulder, as represented by dot *I* in Fig. 19 to the left hand dot at the waist line. Extend the skirt as shown in Fig. 19.

This gives a back with a curve to the arm-hole and a straight line to the shoulder, and is called a combination back. It can be used to advantage for fleshy forms.

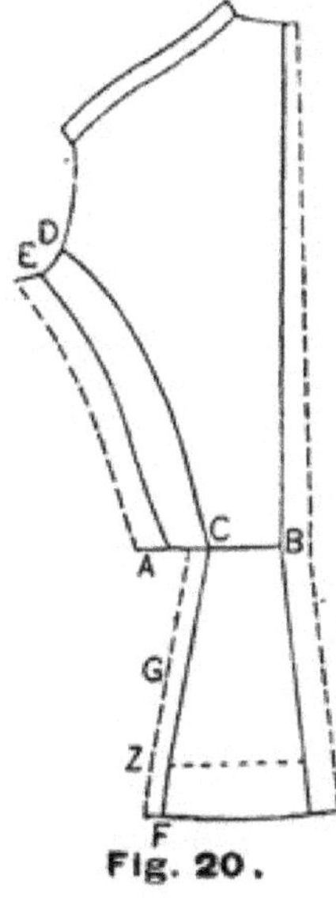

Fig. 20.

# THE FRENCH BASQUE.

### *The French Back.*

This is one where part of the side body is added to the back and a part to the under arm-piece, thus doing away with the regular side seam.

Draft the back and side body, as shown in Fig. 18, so that the sewing lines of the circles of the back and side body will come together at the arm-hole and waist properly. Thus giving us the back and side body together, which makes a back wider than is required. To bring it to the proper width, we decide the width we desired, the back to be at the waist when finished, and place a dot on the waist line at that distance from the centre seam of the back, as at *C*, in fig. 20. The piece of the back to the left of dot, from *C* to *A* is generally about one inch wide. Just as much as you take from the back at the waist line from *C* to *A*, take off from the arm-hole from *E* to *D*. Then with the circle of the side body draw the line *D C*. By placing the waist line of the side body at the waist line *A B*, and the edge of the circle at *D*, a good curve is formed. If you want more curve, drop the waist line of the side body below the line *A B*, and for a less curve raise it above.

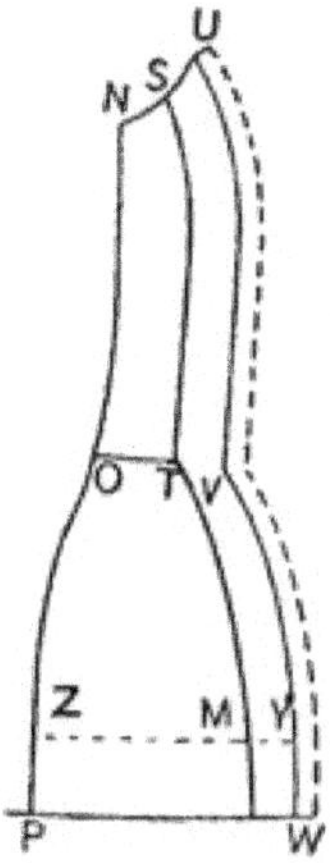

Extend the skirt below the waist by placing the back of the machine as you have it when you drafted the back. Move the machine to the left straight with the waist line until the left edge of the skirt at the waist comes to *C* on the waist line. Mark the left side of the skirt of the back from *C* down to *F*. The skirt requires one-half inch more spring than a plain basque at ZZ, Fig. 20, five inches below the waist. Now the line *A E* is the regular sewing line of the side body, or, in this case, of the sacque back, and *C D* is the sewing line for the French back. The piece between the lines *A E* and *C D* is a part of the waist, but is not used in this back, and must be preserved, as it will be needed to complete the waist. So you can cut this piece out and add it to the side seam of the underarm piece as you will be directed in the instructions that follow for the French underarm piece.

Fig. 21.

### *The French Underarm Piece.*

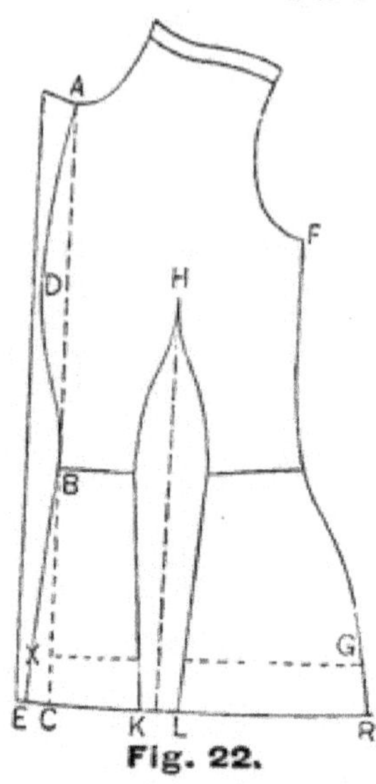


**Fig. 22.**

Mark the underarm piece both sewing and cutting lines and add to the sewing line of the side seam the piece not used in the back ***A E C D*** Fig. 20. This piece of the side body or sacque back that was not used in making the French back needs to be added to the underarm piece to complete it and give the goods required to join the back. It is shown by ***S T, V U,*** Fig. 21. If in making the pattern for the French back you cut the piece off that was not required it is just the size and shape that you need to add to the underarm piece, and by placing its left hand edge at the sewing line of the side seam, then marking along its right hand edge, it will give the sewing line. Allow 1 inch for seam beyond this.

The line ***S T***, Fig. 21, shows the sewing line of the side seam, and from there to the right the piece ***S T U V***, is what is added from the back. Generally this piece is *one inch wide* and its curve is the same as that of the side seam—and as the cutting line of the side seam of the underarm piece is one inch deep, as made by the machine, and has the same curve, we can make the cutting line the sewing line, which adds the piece required, and then add *one inch* for seam. This will give us about the same result as would be obtained by cutting the piece from the back and adding it as above directed, and gives less trouble. Remember, when you *add one inch to the underarm piece* as here directed, you must always take a piece of *the same width off the sacque back* to make the French back. Remember to continue the armhole curve on the piece you have added, as shown from ***S*** to ***U***, Fig. 21. This extra width added to the side seam above the waist is extended below on the skirt.

The skirt is added as follows: The front line is extended straight down for most persons, as shown from ***B*** to ***C***, but those with large stomachs, or above the medium, will require a spring, as shown by line ***B E***. The front darts are extended below the waist the same as in any other basque (see Fig. 22). *The hip measure is applied the same.*

### *The French Front.*

For this front we set the Machine the same as for the regular front as instructed on page 15 or on the diagram, using the curved front. Move the *first dart* over to the left, at top and bottom, thus shutting it out. Bring the lower right hand piece of the *second dart three sizes* to the left, place the top at or near 3½ so as to give the dart a good shape, standing about straight, the top slanting slightly to the right. See Fig. 22. Mark the fold line, neck, shoulder, arm hole, and the underarm seam also the waist line.

Having drafted the front and underarm piece as instructed, we can probably improve the width of the pieces by adding ½ inch to the underarm seam of the front, making it that much wider. This can be done by simply making the cutting line the sewing line and then adding ½ an inch outside for seam. Now as we have made the front ½ an inch wider, we need to take just that much off the underarm piece or our waist would get too large. This is done by simply making the inside or seam line the cutting line, and then marking the seam line ½ an inch to the right, by moving the machine that much.

### *How to Make a Plaited, Gathered or Shirred Waist.*

Draft the lining as for the regular basque, as instructed on page 11, or on the diagram.

Then plait, gather or shirr the outside before cutting and sewing in with or on the lining.

### *To Make a Low Neck*

Arrange the machine and draft as for the regular basque. Cut out such parts about the neck, front and back as you wish removed. Remember, that there is no difference in drafting a low neck from the drafting the plain basque. The only difference consists in cutting out around the neck.

### *To Make a Yoke.*

Draft a plain basque. Then cut out as much as desired around the neck, and cut off front, side body and back at a line about *two* inches below the arm-hole, or shorter, if desired.

---

## THE PRINCESS, POLONAISE, WATER-PROOF, ULSTER, Etc.

### *The Princess.*

Arrange the machine as for a plain basque, as directed on pages 11 to 15, or in the measure book or chart, and add curved front for all forms except those with very small bust. Set the front darts two sizes less for a Princess than for a basque for the same party; thus, if the front darts are marked 12 in the measure book, set at 10 for a Princess. Divide the darts so that the second is a little larger than the first. When the machine is all set, mark all the cutting and sewing lines and the waist line for the front.

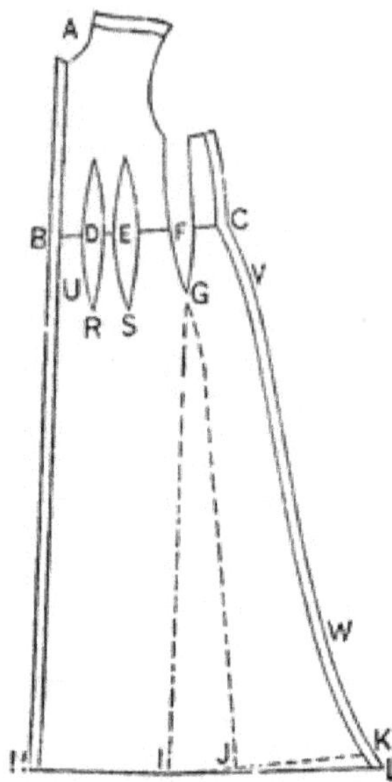

Fig 23.

***To Add the Skirt.***

Extend the fold line straight down from ***B*** at the waist to ***H*** at the bottom, for length of skirt, as shown at figure 23. Add for hem when the bottom is not to be faced. Extend the first dart ***D*** straight down below the waist line 7 inches, and have the dart come to a point there at ***R***. From ***R*** have the lines extend together further down ½ an inch. This will get rid of the fullness at the bottom of the dart.

Make the second dart ***E*** in the same manner.

Extend the undearm seam on the front below the waist, as usual, five inches, by making ***G*** the distance from ***U*** required by the Front Skirt Table on page 17, according to the hip measure. The hip dart ***F*** is formed by the underarm seam line of the front joining the underarm seam of the underarm piece about 5 inches below the waist line at ***G***. The lines extend together at ***G*** ½ an inch or more to get clear of the fullness. The distance the underarm piece is placed to the right of the front at ***F*** can be guided by the spring of the underarm seam. Have the waist line on a line with the waist line of the front, then move to the right or left until the lower end of the sewing line of the underarm seam on this part of the machine comes to the sewing line of the front at ***G***, or you can place the underarm piece to the right of the front as many sizes as the hip is large, according to the measure on the square. A small hip will require about 2 inches space at ***F***, a medium 3, and large 4.

The dotted lines below the hip dart are not often used, only when the goods are very narrow, and then as described hereafter.

The side seam below the waist for line ***C V***, is drafted as usual.

Next decide how wide the bottom of the skirt is to be at the floor line. Is it to be more or less than *three yards?* For a lady that measures 41 inches around the hips, 2½ yards at the floor would be medium. If, then, we wish the garment to be 2½ yards at the floor line, we refer to the table for Princess skirts on page 32, also found in the measure book; find 2½ yards in the first column. Opposite, in the second column, you will find 27 inches, which is the width of the front at the floor, which gives the distance from ***H*** to ***K***, Fig. 23. Measure the length of the skirt at the side from ***F*** at the waist down to ***J***. Draw a line from ***H*** through ***J*** straight to ***K***, for the width of front at the floor.

At ***K***, when the train is very long, add 1½ inches to ***L***, as the train draws the skirt up at this point, and without this allowance makes it short at this place. Have this line, ***C L***, curved in (about 8 inches above the floor line) at ***W***, about 2 inches from a straight line from ***L*** to ***V***. This will cause the train to carry better. This completes the regular Princess front.

The dotted lines from the hip dart down are used only when the goods are very narrow. When the goods are narrow, then make a seam down from the hip dart. *The front line of the hip dart of the waist, springs back regularly to G.* From this on to the bottom at ***J*** it springs back or to the right *one inch* for each *twelve inches* down. If, then, from ***G*** to ***J*** is 36 inches, ***J*** is 3 inches further from the fold line than ***G***. In other words, this line keeps running back slightly from the waist to the floor. The back line of the hip dart is drafted regularly to ***G***, but from this point, where it meets and crosses the front line, it is carried straight down parallel with the fold line in front, or with the grain of the cloth, to the bottom of the skirt at ***I***. When the hip dart is continued down with the seam the front at the bottom extends from ***H*** to ***J***, and the underarm piece from ***I*** to ***L***

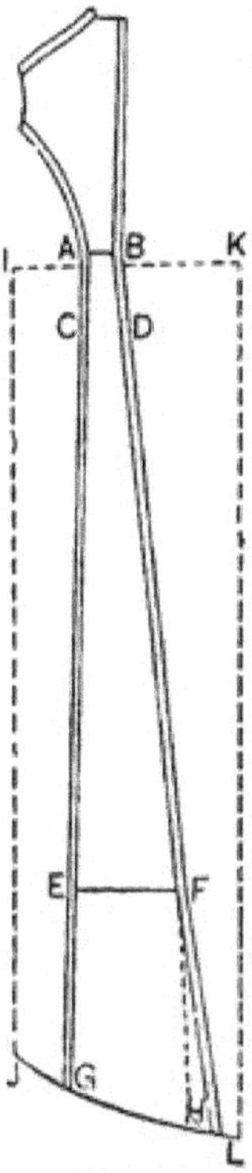

**Fig. 24.**

***How to Make the Back of the Princess.***

Having arranged the machine for the regular back and marked as usual, we have the back made to 5 inches below the waist line. To extend the skirt continue the centre line of the back ***B D***, Fig. 24, to the floor at ***F***, and continue it on to ***H*** until it gives you the length of train desired. For the width of this piece at the floor consult the column for width of back, opposite the width of skirt you are making (in this case 2½ yards) in the table for the princess skirt on page 30, or in the measure book. There you find 9 inches as the width of this piece at this point, the distance from ***F*** to ***E***, at which point make a dot.

Next place the end of the tape at the waist line at ***A***, and at 6 inches down, place it out to the regular spring at the back at ***C***, then continue the line down to ***E***, and extend beyond ***E***, the length desired for train to ***G***. This line curves slightly over the hip just below ***C***. The length of train varies from 2 inches to 2 yards, beyond ***E F*** the floor line of the skirt. The train is shaped after the side body is drafted, and the train part placed as it should go together. Then the edge of ***J L***, Fig. 24, and ***U V***, Fig. 25, can be trimmed to suit the style desired.

When you desire the train to be extra full you can throw extra fullness in the centre seam from ***F*** to ***L***. This is seldom necessary. When plaits or fullness are desired below the waist you can add as shown by the dotted lines ***I J*** and ***K L***, Fig 24.

### *The Side Body.*

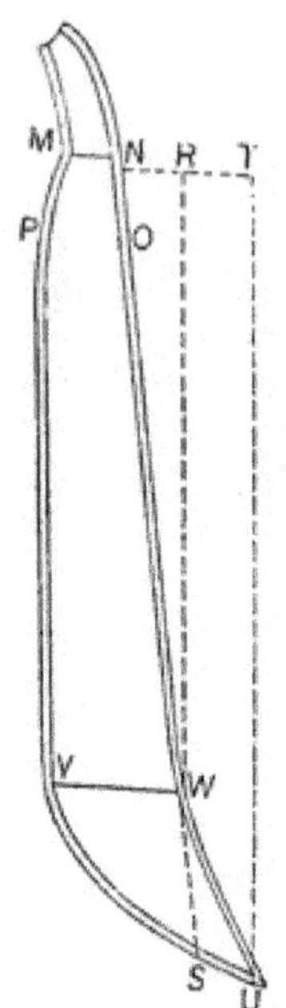


**Fig. 25.**

Having arranged and marked the side body as usual it will complete to 5 inches below the waist line. To extend the skirt continue line ***M P***, Fig. 25, down straight from ***P***, with the grain of the cloth to ***V***, the length of the front on the side. This you can get by measuring. To get the width of the side body at the floor line consult the princess table for skirts. There you will find the distance from ***V*** to ***W*** at which point make a dot. In this case we are using 2¼ yards for the width at the floor and the table gives 11 as the width from ***V*** to ***W***.

Place the end of the tape at ***N*** on the waist line. Spring it out as required to ***O*** five inches below the waist and draw a line from ***O*** to ***W***, the length of the skirt on the back.

To get the train for the side body we continue the line from ***W***, 12 inches to ***S***, and then 4 inches to the right of ***S*** we make a dot. The skirt of the side body which joins the back begins to curve to the right just above ***W***, and passes through the dot to the right of ***S***, which we have just made. The wedge-shaped piece ***S W U*** is to give the spring and allow the train to lay properly on the floor. It is generally added to the side body as we have described, but can be added to the skirt of the back at this point, when the goods are sufficiently wide to permit it to be done without piecing, and it would be necessary to piece the side body.

The dotted line ***R W*** shows how to add fullness or plaits for a short skirt, and dotted lines ***T U*** for a train skirt.

Fashion controls the shape and length of the train, they are all drafted on the above plan, and, from the instructions, you should have no difficulty in following all changes of fashion.

### *Skirt Table, at the Floor For Princess, Waterproof and Ulster.*

| WHOLE WIDTH AT THE FLOOR. | WIDTH OF FRONT AT THE FLOOR. | WIDTH OF SIDE BODY AT THE FLOOR. | WIDTH OF BACK AT THE FLOOR. |
|---|---|---|---|
| 1½ Yards........ | 15 Inches....... | 7 Inches........ | 5 Inches........ |
| 1¾ " ........ | 18 " ........ | 8 " ........ | 6 " ........ |
| 2 " ........ | 21 " ........ | 9 " ........ | 7 " ........ |
| 2¼ " ........ | 24 " ........ | 10 " ........ | 8 " ........ |
| 2½ " ........ | 27 " ........ | 11 " ........ | 9 " ........ |
| 2¾ " ........ | 30 " ........ | 12 " ........ | 10 " ........ |
| 3 " ........ | 33 " ........ | 13 " ........ | 11 " ........ |
| 3¼ " ........ | 36 " ........ | 14 " ........ | 12 " ........ |
| 3½ " ........ | 39 " ........ | 15 " ........ | 13 " ........ |
| 3¾ " ........ | 32 " ........ | 16 " ........ | 14 " ........ |
| 4 " ........ | 45 " ........ | 17 " ........ | 15 " ........ |
| 4¼ " ........ | 48 " ........ | 18 " ........ | 6 " ........ |
| 4½ " ........ | 51 " ........ | 19 " ........ | 17 " ........ |
| 4¾ " ........ | 54 " ........ | 20 " ........ | 18 " ........ |
| 5 " ........ | 57 " ........ | 21 " ........ | 19 " ........ |

## POLONAISE.

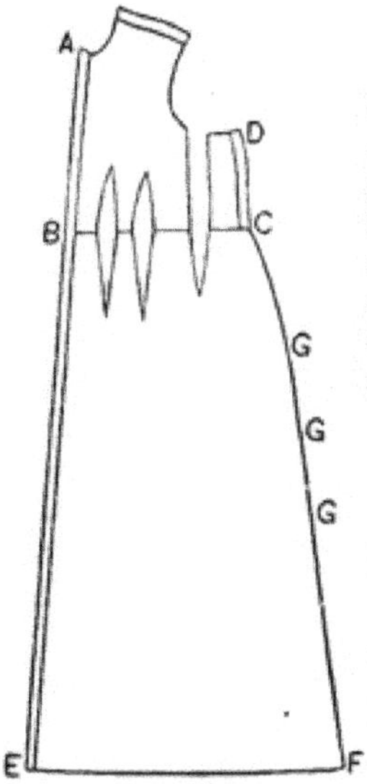

***The Front***

Of the polonaise is made the same as the front of the Princess, just described, except that it is generally two inches shorter and does not curve in above *F*, Fig. 26 as would be required for a train. It is generally draped or fulled along the side at *G G*, as shown in Fig. 26. This front can be changed and draped to coincide with the changes of fashion.

Fig. 26.

***The Back and side Body***

Can be drafted separately as for the Princess, or can be placed about six inches apart on the same waist line as shown by Fig. 27. The circle lines of the back and those of the side body are continued below the waist line, regularly for 6 inches, and then straight on, until they come together at *J*. This makes the skirt of the side body and back in one piece; the centre line down the back is continued from *K* to *R* in a straight line from the waist to within two inches of the floor. The side seam over the hip has been carried down 5 inches below the waist by the machine. As you continue, for the next 3 inches gradually curve into a straight line and continue to the bottom of the skirt with the grain of the goods. If plaits or fullness is desired below the waist they can be added in the centre of the back from *M* to *N*, and between the side body and back by increasing the space between them as desired by *L*.

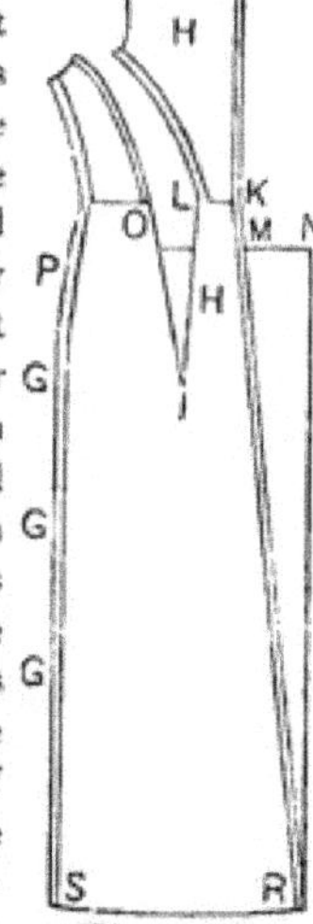

Fig. 27.

The point below the waist at which the fullness is added varies from ½ to 6 inches, according to the fashion. Different drapings will give you as many varieties of styles as you may desire, and this same style of skirt with back and side body together, is used without draping for various kinds of garments.

Princess, Polonaise, Ulster, Waterproof, Duster, &c., are all made on the same general plan, and the same tables and rules apply to them.

## A PLAIN OR GORED SKIRT

Is made with a full breadth of goods in front, one or two pieces on the side, and a full breadth or more in the back. The front and side pieces are gored or tapered at the top. In making this skirt the width of the goods and the size of the figure ought to be taken into account, and from them the number of the pieces and the size of each determined.

A medium size skirt would be 24 inches at the waist, 41 at the hips, and 2 yards and 4 inches at the floor. For half the skirt, at the floor, we would have 38 inches. When the goods are 22 inches wide, the front piece, which is folded, will be 17 inches wide at the bottom from *B* to *C*, *C* is at right angle with line *A B* as shown in No. 1 Fig. 28. The length *A B* is the length of skirt desired, say 38 inches. The top from *A* to *X* you would make 6½ inches wide—about one fourth the waist size. Draw a straight line from ***X*** to ***C*** at the bottom of the skirt. At the top of this line you take out a small gore, 1 inch wide, at the top from ***D*** to ***X***, running out 7 inches below. We also take out a small gore at ***E***, 1½ inches wide at the top and 7 inches deep. This leaves this piece 4 inches at the top when finished.

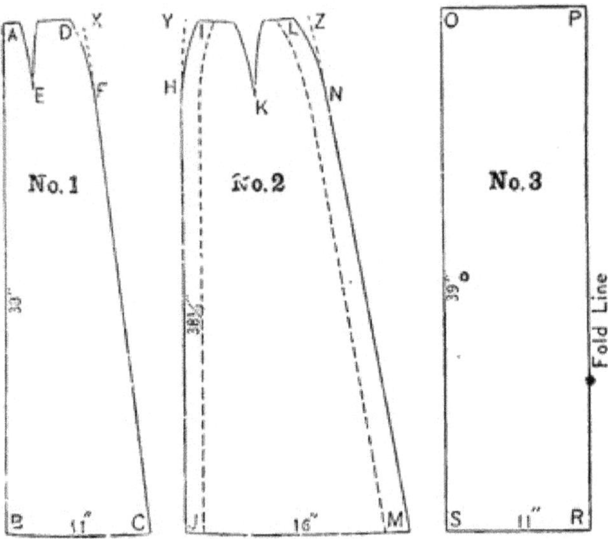


**Fig. 28.**

Next make the piece No. 2 Fig. 28. Make the first line straight from ***Y*** to ***J***, the length of the skirt. Before deciding the width of the bottom of this piece we add the width of the front piece and the width of the back at the floor together, and as they are both folded it will generally be equal to one breadth of the goods; which in this case is 22 inches, which, taken from 38, the half of the skirt at bottom, leaves 16 as the correct width for the bottom of this piece. Place a dot at ***M*** 16 inches from ***J***, on a line at right angles with the front line ***Y J***. At the top, the width from ***Y*** to ***Z***, make a little over one-third the waist size, say 8 inches, and draw line ***Z M***. Take off a small gore from ***H*** to ***I*** on the front line, 1 inch wide at the top, down to ***H***, 7 inches. Also take off a gore from ***L*** to ***N*** 1 inch wide at the top, down 7 inches to ***N***. Also, take out a gore from the centre, 2 inches wide at the top, down 7 inches deep, to ***K***. This will leave this piece about 4 inches wide at the top, when finished, sometimes two breadths are used in the back when that is the case the goods are generally narrowed when two breadths are used you will have one breadth to add to the width of the front-piece.

The back piece is generally a straight piece of full-width goods, and is generally gathered on the band to suit. The back piece can be made gored, the same as the front, but this is not often done. If, for any reason, you should use this gored piece on the back, you will need to run a piece of tape over the seams on the inside, where it joins the side piece, as otherwise both pieces being on the bias, they would sag at the seam. *Remember that whenever two bias edges are united in a seam, you can prevent stretching by covering with a tape.*

When the goods are used 24 inches wide, in place of 22, the front and back pieces will each be one inch wider at the floor, and the side piece can be made narrower in proportion, as shown by the dotted lines.

When the goods are very narrow you can use two full widths in the back or two side gores, as you may desire.

## OUTSIDE GARMENTS, COATS, SACQUES, ETC.

### *The Measure.*

If the party has on a garment such as you wish to make, take the measure over it, just as you would for a basque, making the same *full outside measure* in your measure-book. Set the machine and draft, according to the measures taken. It is, however, so seldom that the party you wish to measure has on such a garment as you desire to make, that it is always better to depend upon the plain basque measure, and then enlarge it for an outside garment. That is, set the machine to the basque measure, and enlarge as follows.

Increase the neck and arm-hole each *one size on both back and front* increase the *width of back length of shoulder each a half size*. Widen the *back* at the *waist two fashion sizes*, and the *skirt* of the *back the same*. Increase the underarm piece one size at the top, at the waist, and at the skirt. Increase the width of the *front length of shoulder one size*, and widen the *front* at the *waist two sizes*. Shorten the lower part of the back and front each ½ inch and wherever the underarm length is used, on the Back, Side body Underarm piece and Front. Change the side body circles to suit the back after they are changed, that is, make the circles correspond with the circles of the back in length, and the side seam with the side seam of the front or underarm piece.

## HOW TO DRAFT A DOUBLE-BREASTED COAT HALF OR THREE-QUARTER TIGHT COAT.

### *The Half Tight Front.*

Having arranged the machine according to the measures, and made the changes just described required for an outside garment, place the front of the machine 4 inches back from the edge of the goods, so as to give the goods required for double breast. Have the waist line the distance from the bottom of the goods or the paper the length of skirt required about 16 inches; with the machine in this position, mark the neck, shoulder, arm-hole underarm seam and waist line. To make the hip dart *D*, Fig. 29, place the underarm piece to the right of the point the space between the two being the number of sizes the hip is large, as found with the square. Have the waist line of the underarm piece on a line with the waist line of

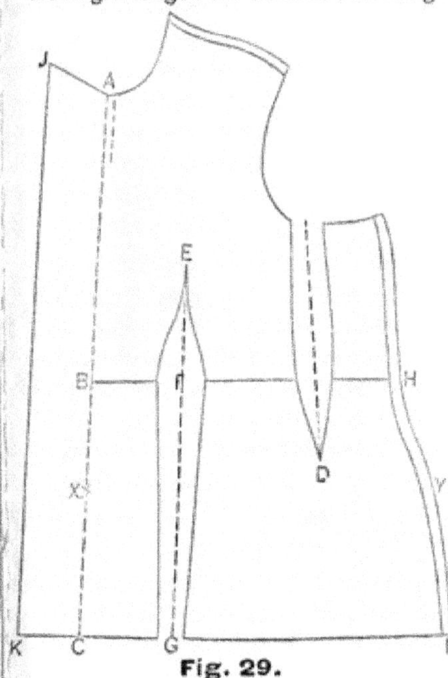

Fig. 29.

the front. Have the underarm seams of the front and underarm piece come to a point at **D** 7 inch below the waist line. *Do not mark the front darts.* Dot down in front of the center line from the neck to the waist, and extend the line on to the bottom of the skirt, as shown by **A B C**, Fig. 29. The center line is the narrow slot running from the neck to the waist line in which the screws and rivets slide. It is ½ inch to the right of the fold line.

At **A**, at the neck of the top of the dotted line, a small gore is sometimes taken out, about ¼ inch wide at the top, running down 4 inches, this is to make the lap keep back in its place and fit properly at the neck. Mark from **A** at the neck out to **J**. This line generally goes straight out; fashion, however, controls this as well as the collar.

The skirt of this coat is made the same as the regular basque skirt, except that the hip seams form a dart, as shown in Fig. 29. When the garment is half tight there will be no darts in front. The above gives us a half tight garment. See back next page.

### *To Make a Three-quarter Tight.*

Make the same as above for a half tight, except that you add *one front dart* as follows: Place the machine as before when you mark the front, then move the machine to the left until the second dart is midway between the center line and the hip dart. Then mark the dart **E F G**, as shown in Fig. 29. See back for this front on next page.

### *To Make a Half Tight Front Without any Darts.*

Simply arrange the machine as before except the hip dart seam, which must be shut out at top and bottom, and not marked. By bringing the underarm piece so its seam line is directly over the underarm seam of the front. Have the waist lines on a line. Allow for an outside garment. Draft the same as before, except the darts. See the back for this front on next page.

### *To Make a Tight-Fitting Front.*

Arrange the machine as for the regular basque; increase the neck one size front and back, increase the width of back, and length of shoulder back each ½ size for thin, and one size for heavy material, increase the width of front 1 size, increase the arm hole 1 size for thin, and 2 sizes for heavy material, both front and back, widen the waist and skirt of the back, each 2 fashion sizes, add 1 size to the waist in front, increase the underarm piece at the top, at the waist and at the skirt each one size shorten the side seam or underarm length of the back, side body underarm piece and front, each ½ inch; change the circle of the side body, so that they will correspond with the circles of the back.

If you want a double-breasted garment place the machine on the goods 4 inches from the edge, and if you want a single-breasted garment place the edge of the machine 1½ inch from the edge of the goods. Have the waist line the distance from the bottom of the goods that you desire the length of the skirt. Mark the center line for double-breasted, or the fold line for single-breasted, neck, shoulder, arm-hole, side seam, waist line, hip dart, and the two front darts. Add the skirt the same as for a basque See back for this front on next page.

## THE BACK.

***Any Side Body For Paletots or Outside Garments.***

Set the back as for the regular basque to the measure taken and make the necessary changes for an outside garment as directed, page 33, that is increase the arm-hole and neck each 1 size. The width of back ½ size, and the length of the shoulder in proportion. Add 2 sizes at the waist and skirt each, and shorten the lower part of the length of back, usually the underarm length ½ inch. Draft as before, except the sewing line in the center of the back. This generally extends straight from the waist to the neck, as shown in Fig. 30.

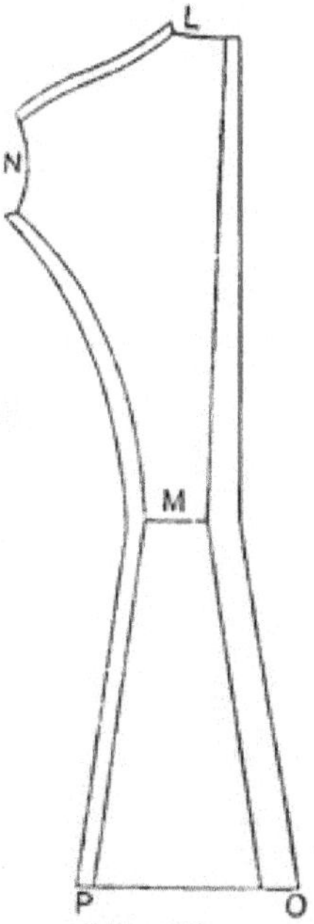

Fig. 30.

***Side Body.***

Set the machine according to the measure taken and make the side seam and circles each ½ inch shorter as required for outside garments. Draft as before and straghten up the side seam below the waist (see Fig. 31), as this garment is a loose one it does not require so great a curve just below the waist. Continue the skirt down to the length desired. This back and side body can be used with either of the fronts, above described on pages 35 and 36.

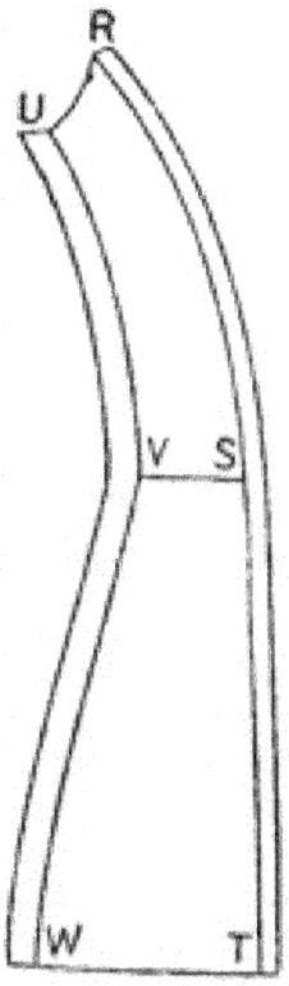

Fig. 31.

***Sacque Back or Back and Side Body Together.***

Set the back as for the regular basque and then increase the neck and arm-hole each one size. Increase the width of back ½ size. Increase the width at waist and skirt 2 fashion sizes. Shorten the lower part of the length of back or underarm length ½ inch, and the side seam and circles of the side body each ½ inch. Mark the back all except the circles. Dot the sewing line of the circles near the arm-hole and near the waist. Place the side body so that the lines at the arm-hole and waist are right, and have the *sewing line* of the circles of the side body over the dots made at the circle of the back near arm-hole and waist. Mark the arm-hole and side seam to the waist, and continue to the bottom of the skirt. Mark the waist line. This back is not often used but can be used with either of the fronts above given. It is the same as Fig. 18, on page 23, except it is changed to an outside garment.

## TO MAKE A FRENCH JACKET.

Take a piece off the Sacque Back about 1 inch wide just inside of the side seam as described on Fig. 20 for a basque on page 25 and add it to the underarm piece, as described and shown in Fig. 21, on page 25. The machine must be changed of course for an outside garment.

## A TIGHT FITTING COAT WITH CROSS SKIRT.

Arrange the machine according to the measure as for the regular basque, and then enlarge it as follows :

*The Back.*

Increase the neck one size, width of back and length of shoulder one-half size for thin, or one size for thick goods. The arm-hole one size for thin or two for thick goods. Widen the waist two fashion sizes and the skirt the same, and shorten the lower part of back length ½ inch.

Fig. 32.

Mark the back as usual all except the seam line down the centre of the back ; this is straighter in coats than basques. Have the line nearly straight from the waist line to the neck, unless the back is very round, in this case you must use the curve to fit. The lap in centre seam of the back is shown by dotted line in Fig. 32. Extend the skirt down the required length say 10 inches.

*The Side Body.*

Simply shorten the circles to correspond with the back as changed, increase the arm-hole one size, and shorten the side seam one-half inch. Mark outside and inside as before, and then cut the skirt off below the waist line the required length 2½ or 3 inches, as shown in Fig. 33, by ***M N O P***.

*The Underarm Piece*

Enlarge the top, waist and skirt, each one size, and shorten the side seam one-half inch, making it the same as the side body or front.

*The Front.*

Enlarge the front as follows : Increase the neck one size, lengthen the shoulder one-half inch to correspond with the shoulder of the back, increase the arm hole one size for thin, or two for thick goods, shorten the side seam one-half inch ; increase the waist one size ; leave the darts as before ; shorten the lower part of the front length one-half inch. Place the outer edge one and one half inch back from the edge of the cloth for single-breasted or four inches back for double-breasted.

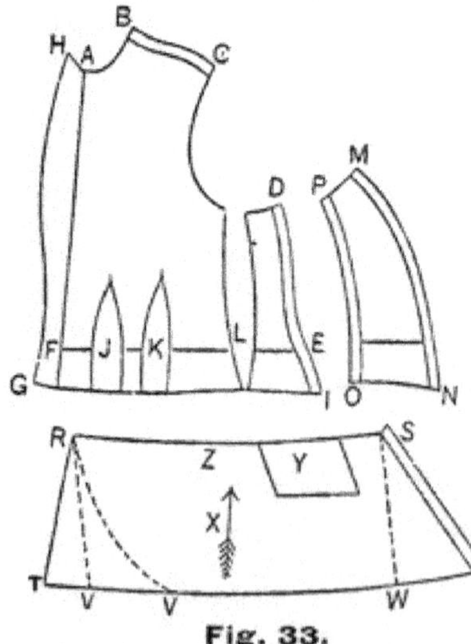


Fig. 33.

Mark the same as usual ; if double breasted mark the centre line ***A F***, Fig. 33, to the right of the fold line ⅓ inch. The front is either straight or curved, as fashion requires. The skirt is added 6 or eight inches long as before and is then cut off below the waist line to suit the fashion from 2 to 4 inches. In this case we will say three inches, as shown by ***G I O N***. The underarm piece and side body are cut off the same length. The line is curved or straight according to fashion ; generally it is straight from the front back to the front darts, then curves up one-half inch at the hip seam or dart and down again at the side seam ; this downward slant is continued across the Side Body Skirt as shown by ***O N***, Fig. 33.

### *The Cross Skirt for Front and Side Body.*

Measure across the skirt of the front underarm piece and side body where you cut it off 3 inches below the waist line from *G* to *N*; don't include the seams and darts. This will give you the width of the front underarm piece and side body when made up. Add two inches to this for lap and it will give the required length of the cross skirt at the top, where it joins the front and side body. Now, if the length of the back skirt is 10 inches long the front will want to be the same. We have below the waist line 3 inches, count ½ inch for seam and we have 2½ finished. The cross skirt then will need to finish 7½ inches; to make the 10 inches to do this it will need ½ inch at top for seam and one or more inches at the bottom, say 9 inches deep. *To draft the cross skirt:* Dot *R* back from the edge of goods 2 inches. Measure straight across to the right to *S*, the length from *G* to *N* as just obtained. Curve this line down in the centre one inch at *Z*, this will give a line to suit the bottom of front, as shown in Fig. 29, but fashion uses curves as well as straight lines across at the top of the cross skirt, that is, adds to the regular skirt at one point, and takes off the cross skirt just as much, so fashion must be your guide here.

The grain of the cloth is shown by *X* the arrow. The depth of the skirt is shown by *R T*, also the spring of the front, when you wish the cloth to meet in front. The lines *R V* represent cut-away fronts, *Y* where to add the pocket, and *S U* the spring where it joins the back. For skirts not over 10 inches long we add half the depth of skirt to get the spring, thus, if it is 8 inches from *S* to *W* it will be 4 from *W* to *U*. This gives plenty of slant; any extra goods at *U* can simply be turned under.

This same plan is used for basques with cross skirts.

## THE COLLAR.

THE STANDING COLLAR is slightly curved. Take the back of the machine and mark a line with the circle the length desired; then move the back from the line just made two and a half inches, and mark another line the same as before. Cut off one inch longer than half the neck size, shape the left end for the front and the right cut square off for the seam, as shown by *A*, Fig. 34.

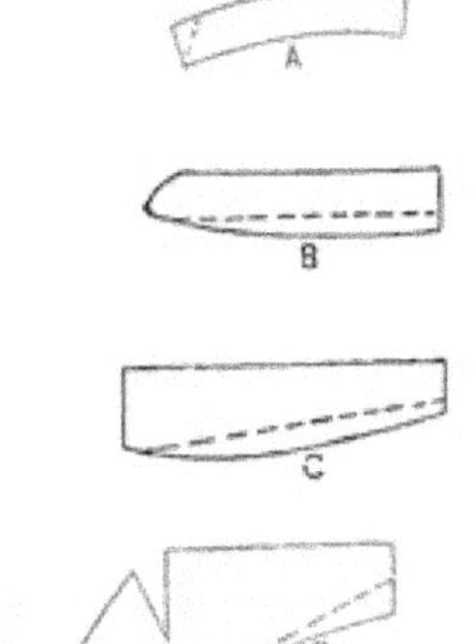


Fig. 34.

THE PLAIN ROLLING COLLAR FOR SINGLE BREAST—see *B*, Fig. 34. It is cut just one inch longer than half the neck size, and one inch wider than it is to be when finished, say three inches. The top is straight, the lower edge slightly curved. The dotted lines show the fold line.

THE WIDE ROLLING COLLAR is made the same as *B*, shown by *C*, Fig. 34. Can be used for double-breasted garments.

THE COLLAR AND LAPEL TOGETHER, is shown by *D*, Fig. 34. It is faced on to the front, which is cut away from the side of the neck to a point five or more inches below the neck on the front line. This will give you the principal styles only.

When A ROLLING COLLAR is used. You can trim out ¼ or ½ inch all round the neck, before putting it in, as they do not require the neck to be as small as the standing collar.

## THE SLEEVE.

One of the most important parts of the dress is the sleeve. When it fits badly it is both uncomfortable and unsightly.

To become a good dressmaker you must thoroughly master this part of your trade. Many a good dressmaker has almost ruined her trade by trying to make one pattern do for all her customers. It would be too short on the shoulder for this one, wrinkle across the full part of the arm for that one, draw across the back for another, bind the arm of the next down, have the elbow in the wrong place, and so on.

*To make a sleeve to fit, and have it tight, requires the following measures;*

| | | | |
|---|---|---|---|
| 1st. | Length of Sleeve. | 4th. | Size at the Wrist. |
| 2nd. | " to the elbow. | 5th. | " of the upper arm. |
| 3rd. | Size of the arm-hole. | 6th. | " at the elbow. |

For each of which there is a place in the measure book.

The size around the arm half way between the shoulder and the elbow, and at the letter point are used simply as test measures to change the regular sleeve, and tighten it up to suit the arm:

*To make a Sleeve to fit medium tight,* four measures are required;

| | | | |
|---|---|---|---|
| 1st. | Length of sleeve. | 3rd. | Length to the elbow |
| 2nd. | Size of arm-hole. | 4th. | Size at the hand. |

We refer to the following table for the information regarding the curve at arm-hole, width of the pieces, etc. And it is to be used in connection with the above measures.

This table gives a curve at the arm-hole suited to the present style, which requires the sleeve to join the body just at the point or joint of the shoulder. Now just as much as you shorten the shoulder, just that much you must add to the curve of the sleeve at the top at 6.

| Sleeve Table for the Arm-hole. | | | | | | From this line to the left this table refers to the Armhole | Elbow part | | | From this line to the right it gives the size at the wrist | Table for sleeve at wrist | | | | | | |
|---|---|---|---|---|---|---|---|---|---|---|---|---|---|---|---|---|---|
| Size of Arm-Hole. | 4 Front Curve from 1 to 4 | 5 Back Curve from 1 to 5 | 6 High Point from 1 to 6 | 7 Width of Upper from 5 to 7 | 8 Width of Under from 5 to 8 | | 9 | 10 | 11 | | 12 | 13 | T | S | Size at wrist Finished. | 14 Width of upper Piece from 12 to 14 | 15 Width of Upper Piece from 12 to 15 |
| 10 | 2½ | ⅞ | 3½ | 7 | 4 | | The length to the Elbow from 7 to 9 | Always 2 inches in from 9. | Always 2 inches in from 3. | | Always 2 inches from 2. | Always ½ inch from 12. | Always 2 inches above 13. | Always 1½ ins. to right of T | 7¼ | 5⅛ | 4⅛ |
| 11 | 3 | 1 | 4 | 7½ | 4½ | | | | | | | | | | 7½ | 5¼ | 4¼ |
| 12 | 3¼ | 1⅛ | 4½ | 8 | 5 | | | | | | | | | | 7¾ | 5⅜ | 4⅜ |
| 13 | 3½ | 1¼ | 4¾ | 8½ | 5½ | | | | | | | | | | 8 | 5½ | 4½ |
| 14 | 3¾ | 1⅜ | 5 | 9 | 6 | | | | | | | | | | 8¼ | 5⅝ | 4⅝ |
| 15 | 4 | 1½ | 5¼ | 9½ | 6½ | | | | | | | | | | 8½ | 5¾ | 4¾ |
| 16 | 4¼ | 1⅝ | 5⅝ | 10 | 7 | | | | | | | | | | 8¾ | 5⅞ | 4⅞ |
| 17 | 4½ | 1¾ | 6 | 10½ | 7½ | | | | | | | | | | 9 | 6 | 5 |
| 18 | 4¾ | 1⅞ | 6⅜ | 11 | 8 | | | | | | | | | | 9¼ | 6⅛ | 5⅛ |
| 19 | 5 | 2 | 6⅝ | 11½ | 8½ | | | | | | | | | | 9½ | 6¼ | 5¼ |
| 20 | 5¼ | 2⅛ | 7 | 12 | 9 | | | | | | | | | | 9¾ | 6⅜ | 5⅜ |
| 21 | 5½ | 2¼ | 7⅛ | 12½ | 9½ | | | | | | | | | | 10 | 6½ | 5½ |

This table can be used for the arm-hole and wrist separately as two different tables, using the part to the left for the arm-hole table, and the part to the right for the wrist table. Or it can be used as one table by following the line through both parts of the table for the arm-hole size. This will give a proportional size of elbow and wrist to the arm-hole.

We can also get a proportional length to the elbow for a sleeve, by dividing the length into equal parts, then adding 1 inch to the upper half and taking 1 inch off the lower half. Thus, sleeve 22, one-half, 11; 1 inch added for upper length makes 12 to elbow and 10 for length from elbow to wrist.

A plain or proportional sleeve can thus be drafted from the size of arm-hole and length of sleeve, but you will always find it best to get the size at the wrist and to locate the elbow point, for any sleeve which you desire to have fit closely.

In Fig. 35, we use figures to designate the order of making or learning to draft the sleeve. The figures, 1, 2, 3, &c., show the order of the things to be done. The distance from 1 to 2 the length of sleeve, from 1 to 3 the length to the elbow. The size of arm-hole, and size at hand are found in the measures, and the figures 1, 2, and 3 are located by them, all the other figures are located by table on page 38. In other words you can find in table all the lengths, without figuring, needed for the sleeve that are not found in the measures themselves.

To use the table find the arm-hole size in the first column, then follow out the line to the right, in the second column which has 4 at the top you will get the distance from 1 to 4, in the next column from 1 to 5, and so on through. In the table you get the distance and in Fig. 35 the location for each point or figure. Remember all the distances not given in the measures are found in the table.

### *To Draft the Sleeve.*

We follow the figures in learning, 1 is the first dot or starting point, 2 the second dot, and so on.

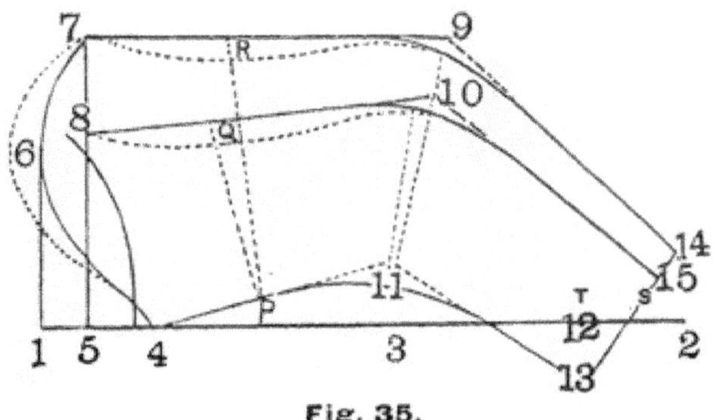


**Fig. 35.**

These figures are taken from the measures or the table.

1st. Get the length for sleeve from the measures.

2nd. Draw line from 1 to 2 full length as shown in Fig. 35 *in this case* .......... 21

3rd. Dot at 3, the length to the elbow which is from 1 to 3 and is found in the measures, *in this case* .......... 12

4th. Dot at 4 the length for the inside curve at the arm-hole. This is found in the table. Find the arm hole size, in this case 14, in the first column, follow this line on through the table to the right, in the second column, which has 4 at the top, you will find the proper distance from 1 to 4, *in this case* .......... 3¾

5th. Dot at 5, the length for the back curve at the top, which is found in the table in the third column, which is marked 5 at the top, and is the distance from 1 to 5, *in this case* .......... 1¼

6th. Draw line from 1 to 6 the distance for the high-point. This is found in the table in the fourth column which has 6 at the top. The high point or 6 is out from 1 at a right angle to line 1 to 2. The distance *in this case* .......... 5

7th. The width for the upper part of the sleeve at the arm-hole from 5 to 7 is found in the table in the column which has 7 at the top, draw line from 5 to 7 at a right angle with line 1 to 2 *in this case* .......................................... 9

8th. Dot at 8 the width for the under part at the top, the distance from 5 to 8 is found in the table in the column which has 8 at the top, and *in this case* .......................................... 6

9th. The distance from 7 to 9 is the length to the elbow as found in the measure, draw line from 7 to 9 square with line 5 to *in this case* .......................................... 1

10th. The distance from 9 to 10 is the difference in the width of the pieces at the elbow and is always 2 inch in the draft, it is often changed after the draft is made, draw line from 10 to 8 ....... 2

11th. The distance from 3 to 11 for the inside curve is always 2 inches, draw line from 11 to 4 .......................................... 2

12th. The distance from 2 to 12 is always 2 inches and is the end of the sleeve at the hand .......................................... 2

13th. The distance from 12 to 13 controls the curve of the sleeve it is usually ½ inch but varies some with fashion, when more than ½ inch is used the sleeve curves more, the dot 13 should be directly opposite and below the 12 outside the line; draw line from 11 to 13 .......................................... ½

14th. Directly above 13 two inches dot at T and then dot at S 1½ inches to the right this gives the angle for the line at the hand from 13 through S, the distance from 13 to 14 is found in the table. Find the size at the hand in this case 8 in its column and on a line to the right in column with 14 at the top you will find the distance to 14, draw line from 13 to 14 *in this case* 5½

15th. The distance from 13 to 15, the width of the under piece at the hand, is found in the table on the line to the right of the size at the hand in the last column which is marked 15 at the top and is .......................................... 4½

Draw lines from 9 to 14 and from 10 to 16. Curve at 9, 10, and 11 with the circles of the back or side body.

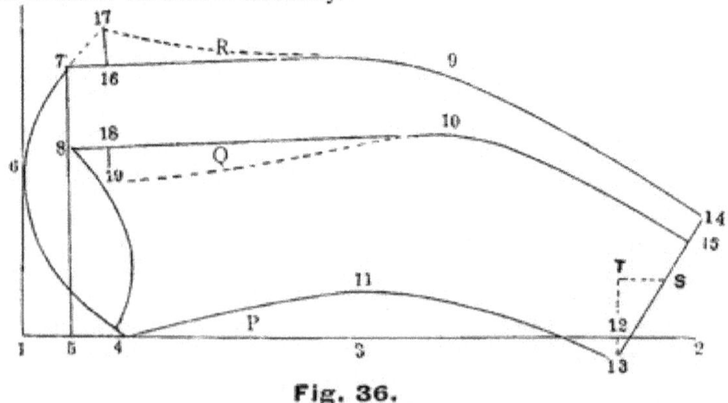


**Fig. 36.**

***To Make a Sleeve with a wide Upper and Narrow Under.***

Draft the sleeve just described and shown in Fig. 35, page 39 according to the measures and the table as explained. Then dot, as shown in Fig. 36 1½ inches from 7, on line 7 to 9 at 16 and dot at 17 out square from

16, the same distance 1½ inches. Extend the armhole from 7 to 17 and draw a line from 17 to 9 we have thus added a piece 1½ inches wide to the upper piece at the armhole. To keep the sleeve the same size take off of the underpiece a piece as large as was added to the upper to do this dot at 18 on line 8 to 10 in from 8, the 1½ inch, then dot at 19 in on the underpiece 1½ inch in from the 18, draw a line from 18 to 10. At the armhole part of the under armpiece a half inch is added for seam.

You can add as much as 2 inches to the upper piece and make the under that much less by making the distance from 7 to 16, and 16 to 17 each 2 inch, and the distance from 8 to 18, and 18 to 19 the same.

***To get the Curve of the underpiece at the armhole use the side body.***

The curve for the upper piece at the armhole is from 7 to 6 and then down to 4; have it broad *F* and not sharp. The better plan is to make a dot about 1½ inches to the right and 1½ to the left and ¼ inch above at 6, then extend this line down to 8 and down to 4. Have the line at 6 almost straight for the 3 inches.

*To make the sleeve tight fitting*, use the test lines *P Q* and *P R* for the upper part, and 11, 9 and 11, 10 at the elbow. That is, measure the sleeve at these points, and if they exceed the measure you have in the measure book more than two inches, take out the extra goods at *R* and *Q* for the upper part, and 9 and 10 at the elbow. Shape as shown by the dotted lines.

*Extra fullness on top of the shoulder or a puff* is made by the curve of the upper piece beyond 6 Fig. 35.

*For a very short short shoulder* the top of the sleeve must be extended at this point. Add to the curve as you shorten the shoulder.

*A straight or curved sleeve* can be made by simply changing the distance from 12 to 13. When 13 is ½ inch from 12, the sleeve curves about right for a common sleeve, and when it is one inch, it has considerable of curve.

*To make a slight fullness* at the elbow, draft as instructed above. To get clear of the fullness at the elbow add ¼ inch to the underpiece from 10 to 15 and take ¼ inch off the upper piece from 9 to 14. Or leave sleeve as first drafted and add ¼ inch on or under at 15. Remember there is always a half inch added to the armhole part of the underpiece from 4 to 8. When the upper piece is some longer than the under, the extra length is fulled in at the elbow in a space of two inches.

Draft as many as 6 as sleeves of this style and that will fix the figures in your mind, so you will know just where they belong without marking them. Keep at the sleeve until you master it.

*Cut out the pattern on the lines as is marked*, except at 6, where you must allow extra for short shoulder or puff sleeve. The goods for seams are included in the draft and are not to be allowed extra. The seam must be taken a full half inch deep, or they will be too large.

*To baste the sleeve*, start at the arm-hole and baste the inside seam first. Then put the back seam together at the top and baste evenly up to within two inches of the elbow point: next start the wrist and baste up to within two inches of the elbow. Gather the upper part of the sleeve at the elbow on a thread, then baste it to the under part. This will bring whatever fullness there may be in the sleeve in its proper place.

*To put the sleeve in*, start the back seam, at the arm-hole, at the top of the curve where the side body joins the back, this is nearly correct for the sleeve as drafted from Fig. 35, but when changed as instructed in Fig. 36, the back seam comes near the top of the side seam, a half inch seam is taken in the arm-hole in both the sleeve and the body; baste evenly up to the shoulder seam; from there up over the shoulder full the sleeve. The highest point of the sleeve goes to the highest point of the shoulder. Putting in the sleeve at the arm-hole is very important, as the sleeve will twist if not right. Remember the deep seam.

### TO MAKE A SLEEVE WITHOUT THE BACK SEAM FROM THE ELBOW TO THE SHOULDER.

Draft the sleeve as shown in Fig. 35, page 41. Take off the under piece with the wheel, then lay the upper and under pieces with the two back edges together (lines 7 to 9 and 8 to 10), Fig. 35, so they lap one inch. Have them so the arm-hole continues smoothly at the top. Mark around the outer edge and the gore at the elbow.

FOR A SHORT SLEEVE, cut off 3 or 4 inches below the elbow.

*Remember, in putting the sleeve in, that the highest point of the sleeve goes to the highest part of the arm-hole.*

*Remember that half inch seams are included in these drafts.*

*Remember that coat sleeves do not curve quite as much as basque sleeves.*

*Remember that the following instructions enable you to make a sleeve with any curve desired, by simply changing the distance from* 13 *to* 12.

## CIRCULARS AND CAPES.

These wraps are cut in a variety of styles and shapes, but are all made on the following simple plan. The front and back of the basque are placed so as to form the shoulder seam and give the goods required over the arm : Set the machine as you would for a basque, increase the neck and width of the back each one size and the neck of the front one size, make the shoulder of the front the same length as the shoulder of the back.

Place the edge of the front of the machine back 3 inches from the edge of the goods and nine inches from the top, as shown by distance ***A O***, Fig. 37.

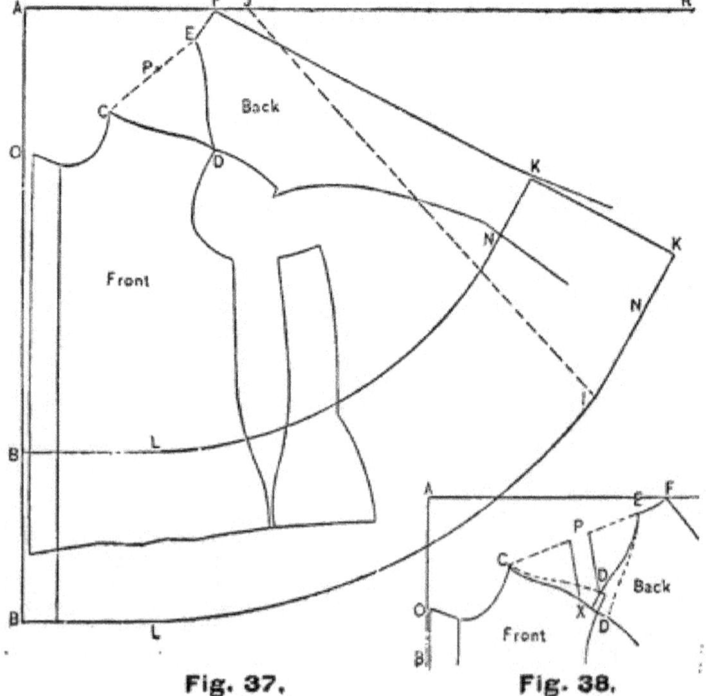


Fig. 37. Fig. 38.

Next place the back so the seam lines of the shoulders of the front and back come together at the arm-hole at *D* Fig. 37. With this point as a pivot move the back, until the center line *F K* has the direction required. The angle here given to the line *F K* gives you the medium amount of fullness in skirt. If it is moved to the dotted line *I J*, which is nearly a full bias, it will give less goods and make a close fitting skirt. To make a close or loose hanging garment, simply change the position of the center line of the back.

To get the length, measure the front or back which ever you have, then draw a line straight in from *B* or *K* the distance *P* is from the center line of the front or back. Next place the end of the tape at the center of the shoulder gore at *P* and make a sweep from *N* to *L* for lower edge.

*When you want the seam on top of the shoulder* add one and a half inches (not sizes) to the arm-hole of the back, that is raise the shoulder seam at this point that amount. And take off the arm-hole in the front the same amount added to the arm-hole of the back, as shown in Fig. 38. The shoulder lines come together then at *X* in place of *D* as in Fig. 37. The back and front are placed the same, and the skirt made the same.

### SIDE SEAMS CAN BE USED IN PLACE OF THE ONE DOWN THE CENTRE OF THE BACK.

When this is desired the pattern is made the same as Fig. 38, but the goods are folded on the center line of the back and the shoulder seam continued over the shoulder straight down to the bottom.

To HAVE TWO SHOULDER GORES, simply have a strip of goods one and a half inch wide in the center line of the back and the large gore from *P* to *X* as shown in Fig 38. This gives you two small gores *C X P* and *P D E*, you must allow for seams.

These illustrations will give you the general plan, and a little practice will enable you to follow any fashion.

## THE DOLMAN

Has a variety of styles and can be produced in several ways. The simplest, and perhaps the best, plan is to set the machine according to the regular basque measure, increase the neck one size and draft the front and back, then change these, as required to suit the different styles and add the sleeve to suit the fashion.

To assist in making these changes it is well to have the following

### *Test Measures.*

1ST MEASURE.—With the arm bent at a right angle, and the hand resting upon the pit of the stomach with the arm in an easy position at the side, measure from the center of the back, at the height of elbow around over the arm to the little bone at the wrist; add *two inches* for seams and comfort.

2D MEASURE.—Measure from the center of the back from a point *six inches* below the neck straight around over the arm to the arm-hole seam in front. Add *two inches* for seams and comfort.

### *Draft the Front as Follows.*

Set the machine according to the regular measure. Increase the neck one size.

Place the edge of the machine even with the edge of the goods and have the waist line above the lower edge the required length of skirt. Then mark the fold line, neck, shoulder, arm-hole, underarm seam and waist line. Do not mark the darts, but dot at the underarm seam at the waist line.

Extend the fold line down below the waist, as shown in Fig. 39, to the length desired.

Take half the full hip measure, and apply it across the skirt *six inches* below the waist line, from **Z** on the fold line to **Z** on line **F G**. Add *one inch* to complete the length; that is, use ½ the hip measure plus *one inch*. Thus, if the hip measure is 42, the ½ is 21, to which add one inch, we then have 22 inches, the length required from **Z** to **Z**.

Have the line **F G** spring to the right at the rate of *two inches* for each *six inches* it is continued below **Z**; thus if **G** is 12 inches below **Z**, then the line **B G** will extend 4 inches further to the right than the line Z Z. When the Dolman is very long, 1 inch will be spring enough in 6.

The width of the skirt of the front as here given is based on the usual width of the skirt of the back, which is 4 inches wide at 6 inches below the waist line from **Z** to **Z**, Fig. 40. Just as much as the width of the back at this point *exceeds 4 inches*, you will deduct that amount from the width of the skirt in front on line **Z Z**, making it that much narrower.

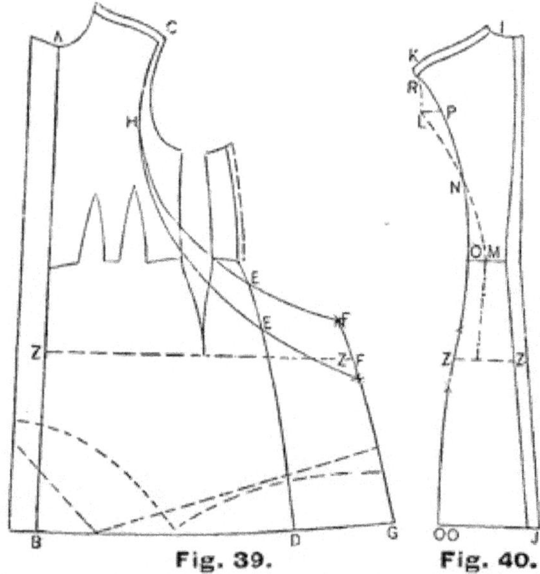


**Fig. 39.** **Fig. 40.**

To finish the front we cut away that part or portion of the front not required for the Dolman. Begin at the shoulder at **C**, Fig. 39, at half inch inside the arm-hole and continue that distance inside the arm-hole on down to **H**, which is a little above the lower part of the arm-hole. From **H** spring back to the right and cross the waist line three inches to the right of the underarm seam, and extend it to **F**, 2 inches above **Z**. When a lower cut is desired, strike **F** below **Z** by crossing the waist line near the under-arm seam in place of the right as shown in Fig. 39.

### *The Back.*

Place the back on the paper so there will be room to add the sleeve at the left. Then mark all the back except the arm-hole and circles. Then dot these on the sewing lines as shown in Fig. 40. Next dot at *K* on the shoulder half inch from the arm-hole and at *P*, 1½ inches inside of *L*, the sewing line of the arm-hole, and at *N*, the point where the curved line crosses the circles about ⅔ of the distance on the circle from the waist up, also at *O*, at the waist, which is 2 inches to the left of *M*, the width of the back at the waist line.

Draw the curved line from *K*, through *P* and *N* to *O*. See that this line has a graceful curve. Extend the skirt on down, the length desired, keeping 1½ inches to the left of the regular basque skirt, 6 inches below the waist line at Z and on to *O O*. This line curves slightly from the waist to the bottom. Finish the back by drawing a line across the bottom of the skirt as shown in Fig. 40.

### *How to Get the Quantity of Goods Required.*

This table gives the quantity of material required to make the following garments, medium sizes, for different width of goods.

The figures on the lines opposite the name of garment, is the number of yards required to make it. At the top of each column is marked the width of the goods. So to get the quantity of material for different widths you have only to look in the different columns.

To use the table find the style of garment in the first column, go out to the right to the column marked with the width of goods you want and there is the amount required.

The table is based on 36 inches bust measure, and medium figure, and will be near enough right to be used for 34, 36, 38 without change. When, however, there is much deviation in size from the 36 bust, you can add or take off a little as your judgment suggests.

### *The Sleeve.*

After drafting the back as just described, and shown in Fig. 41, take a plain basque sleeve *to suit the arm size* and place the top of the back seam

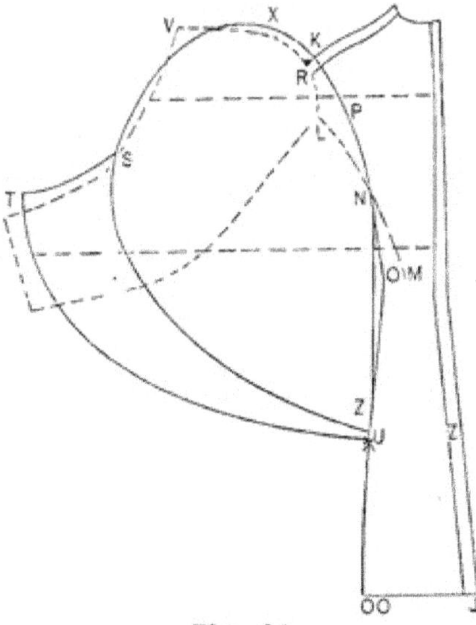


Fig. 41

at *L*, the arm hole of the back, Fig. 41, and stick a pin at this point. Then with this point as a pivot move the sleeve so the curve of the arm-hole of the sleeve joins the curve of the arm-hole of the back at *R*, on the dotted line.

With the sleeve in this position, dot all around it. Then stick a pen at *S*, the elbow of the inside curve of the sleeve, and take the one at *L* out and using the pin at *S*, as a pivot move that part of the sleeve at the wrist up to *T*, locating *T* at about the same height as *S*, this will make the forepart of the sleeve nearly straight out. Then mark from *T* to *S*, and continue toward V, to within 3 inches thereof, keeping a slight curve at *S*. From this point continue the curve on to *K*, keeping inside of *V*, about 1 inch, and outside of *X*, ½ inch. That is, we take off the point at *V* and extend the curve at *X* beyond the regular sleeve curve ½ inch, or as much as we shortened the shoulder at *K*. In other words, we must add to the sleeve at this point as much as we take off at the shoulder.

That part of the sleeve which joins the back has the same curve as the back from *K*, through *P* to *N*, from which point the sleeve continues on down in a straight line to *Z*, 6 inches below the waist line, and then on to *U* the length desired. It generally continues on down the back to a point 2 inches below where the skirt of the front joins the back. The shape of the sleeve from *T* to *U* varies according to fashion. Sometimes it is a regular curve from *T* to *U*, at others it is carried straight down from *T* and straight across from *U*.

The under part of this sleeve is the same from *T* to *S*. From *S* it can be a regular curve to *U*, the same as the curve of the front from *H* to *F*, Fig. 39.

*In sewing in the sleeve, the highest point of the sleeve goes to the highest point of the arm-hole. This brings point **S** of the sleeve near point **H** on the front.*

From *H* to *F* on the front, Fig. 39, and from *S* to *U* on the under part of the sleeve, Fig. 41, are sewed together, and a piece of tape stitched over the seam.

The under part of the sleeve need not extend all the way from *S* to *U*. It can stop ½ of the way if desired. The lower edge of the under piece is the same as the lower edge of the outside piece, and completes the sleeve as shown in Fig. 41. The lower edge of this sleeve is open.

Apply the test measures first from the center of the back to *T*, and then the second measure from the center of the back through R to a point a little above *S*. This completes the open sleeve.

WHEN YOU DESIRE A SLEEVE CLOSED ALONG THE LOWER EDGE.

The front and back are drafted just the same, and the sleeve is made the same, but is shaped differently at the bottom, as is seen in Fig. 42. Draw a line straight out to the left from *U* to *Y*, the lower edge of the sleeve, and a line straight down from *T* to *Y*, this gives the lower edge of the outside piece. To get the under part of the sleeve fold the paper on line *YZ* across the bottom, shape the end from *Y* to *T*, and *T* to *S* the same as the outside curve; from *S* to *U* curve the same as the front. This makes the sleeve all in one piece as the goods are folded on the lower edge thus saving the seam there. The test measure can be applied the same as before.

This is the favorite sleeve. It can be left full size at the band or gathered, as fashion requires.

## TO HAVE THE BACK AND SLEEVE ALL IN ONE PIECE.

Make the front the same as before shown, Fig. 39, and described on page 44.

Make the back the same as before shown, Fig. 40, and described on page 44.

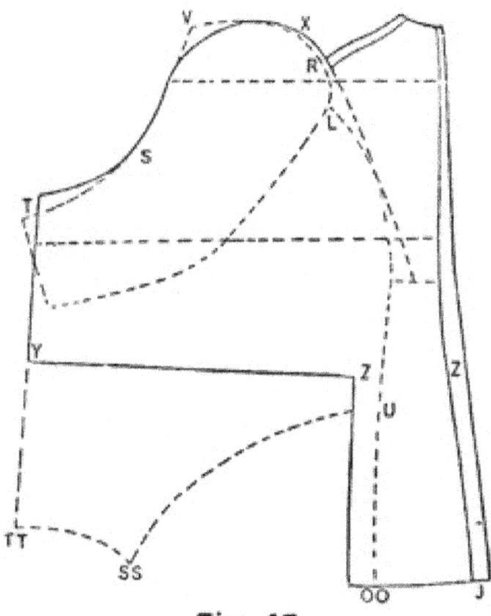


**Fig. 42.**

The sleeve is added just the same way but the back and sleeve are cut all in one piece, that is the seam joining the sleeve to the back is done away with.

Cut the sleeve to suit the fashion. The closed sleeve just described and shown in Fig. 42 is very suitable for this style of back and sleeve in one.

These samples will serve to give you a good idea of the Dolman, but to master it will require some practice as it is considered by some the most difficult garment to make.

## HOW TO FIND THE QUANTITY OF MATERIAL IN A MADE UP GARMENT.

Find the number of square inches in each piece of the garment and add them together. This will give the number of square inches; to get the number of yards, divide this amount by the number of square inches in a yard of goods the width you require.

To do this, take the average width and length of each piece in inches Multiply them together and you will get the square inches in each piece. Add the inches in the several pieces together.

When you measure plaits, folds or ruffles made up, take three times the square inches they contain, as they will require three times the amount of goods to make them up.

In reducing the total amount of square inches to yards, always multiply 36 inches by the width of the goods. If the material is 20 inches wide you have 36x20—720 in each yard. And if the whole number of inches were 10,800, by dividing that by 720, gives 15 yards; one yard in every 10 is added for loss, which added to 15, makes 16½ yards of 20 inch goods. You can make a very close measure with a little care.

## Amount of Material for Different Garments.

| Style of Garment. | Different Width of Goods in Inches. | | | | | | |
|---|---|---|---|---|---|---|---|
| | 18 | 20 | 22 | 24 | 27 | 36 | 48 |
| | Yds. | Yds. | Yds. | Yds | Yds. | Yds | Yds. |
| Belt Waist | 2½ | 2¼ | 2 | 1⅞ | 1⅓ | 1⅓ | 1 |
| Sack " | 5 | 4⅛ | 4 | 3⅛ | 3⅓ | 2½ | 1½ |
| Basque with 9 in skirt | 6⅛ | 5⅓ | 5 | 4¼ | 4 | 3 | 2⅓ |
| " 12 " | 6⅓ | 5½ | 5⅛ | 4⅝ | 4⅛ | 3¾ | 2¼ |
| Polonaise, short | 8⅓ | 7¾ | 7 | 6¼ | 5⅔ | 4⅓ | 3¼ |
| " medium | 9¾ | 8⅛ | 8 | 8⅓ | 6⅓ | 4⅞ | 3⅔ |
| Princess Wrapper with 6 in train | 12¼ | 11 | 10 | 9¼ | 8 | 6 | 4¼ |
| " " " 10 " | 15¼ | 13¾ | 12½ | 11¼ | 10¼ | 7½ | 5¼ |
| A Princess with full train | 19⅝ | 17⅓ | 16 | 14⅔ | 13 | 9⅔ | 7⅓ |
| Coat, medium | 5¼ | 5 | 4⅓ | 4 | 3⅔ | 2¾ | 2 |
| " long double breasted | 6⅜ | 6 | 5⅓ | 4⅝ | 4⅓ | 3¼ | 2⅔ |
| Cape or Wrap, short | 3 | 2¾ | 2⅜ | 2¼ | 2 | 1⅝ | 1⅓ |
| Wrap, medium | 6⅛ | 5⅝ | 5 | 4⅛ | 4 | 3 | 2¼ |
| Cloak or Long Wrap | 8⅓ | 7¾ | 7 | 6¼ | 5¾ | 4⅓ | 3¼ |
| Dolman, short | 5 | 4⅝ | 4 | 3½ | 3⅛ | 2⅛ | 1⅝ |
| " medium | 5¼ | 5 | 4½ | 4 | 3⅞ | 3 | 2⅔ |
| " long | 6½ | 5½ | 5 | 4⅛ | 4 | 3 | 2¼ |
| Suit, plain | 14⅜ | 13¼ | 12 | 11 | 9⅞ | 7⅛ | 5⅛ |
| " medium | 17⅛ | 15⅓ | 14 | 12⅝ | 11⅓ | 8⅓ | 6¼ |
| " fancy | 19⅝ | 17⅓ | 16 | 14⅔ | 13 | 9⅛ | 7⅓ |
| Skirt, plain | 9⅜ | 8⅛ | 8 | 7⅛ | 6⅓ | 4⅞ | 3⅔ |
| " fancy | 12¼ | 11 | 10 | 8¼ | 8 | 6 | 4⅛ |
| Overskirt | 7⅓ | 6¾ | 6 | 5⅓ | 5 | 3⅔ | 2¾ |
| Sleeve, plain | 1¼ | 1⅛ | 1 | 1 | ⅞ | ⅔ | ⅓ |
| " fancy | 2½ | 2¼ | 2 | 1¾ | 1⅓ | 1⅓ | 1 |

### *Girls and Misses.*

A girl of 7 years of age, requires

For plain costume................4 yards of 22 inch goods.<br>
" fancy " ................7½ " " "<br>
" cloak " ................3 " " "

A miss 14 years of age, requires

For Ulster......................2½ yards of goods 48 inch.<br>
" medium costume............8 " " 22 "<br>
" fancy " ............10 " " 22 "

Plaiting, ruffling and shirring, as a rule, require three times the length you desire them to be when finished. In other words three times the quantity of yards as when made plain.

# INDEX.

www.ingramcontent.com/pod-product-compliance
Lightning Source LLC
Chambersburg PA
CBHW022041080426
42733CB00007B/930

* 9 7 8 3 3 3 7 3 1 4 2 2 4 *